1001 CHEMICALS IN EVERYDAY PRODUCTS

1001 CHEMICALS IN EVERYDAY PRODUCTS

SECOND EDITION

Grace Ross Lewis

A Wiley-Interscience Publication

JOHN WILEY & SONS, INC.

New York • Chichester • Weinheim • Brisbane • Singapore • Toronto

Copyright © 1999 by John Wiley & Sons, Inc. All rights reserved.

Published simultaneously in Canada.

Library of Congress Cataloging-in-Publication Data:

Lewis, Grace Ross.
    1001 chemicals in everyday products / Grace Ross Lewis.—2nd
  ed.
      p. cm.
    "A Wiley-Interscience publication."
    Includes index.
    ISBN 0-471-29212-5 (pbk. : alk. paper)
    1. Chemicals. I. Title.
  TP200.L49 1999
  363.17'9—dc21   98-6419

Printed in the United States of America.

10 9 8 7 6 5

To Dick
for a million and one reasons.
Special thanks to Richard J. Lewis, Jr. for his technical computer advice, creative ideas, and interest in this project. Many thanks to Julie R. Lewis, Esq. for materials and inspiration. Thanks to Velma W. Ross for her kind thoughts.

CONTENTS

PREFACE

This book is designed for you, the consumer. Whether knowledgeable about chemicals or not, everyone will find it an informative source of data. It is different in that it is a book about chemicals, but it is not a chemistry book in the traditional sense. This book has a different purpose: that is, to answer your questions and provide information about all the chemicals you come in contact with every day.

In the past we have done research for chemical reference books for professionals in the environmental and occupational health field. In the course of doing research for books about workers' occupational exposure to chemicals, we became curious about how many of these chemicals we consumers contact in our lives and homes as ingredients in foods, cosmetics, and cleaning products.

Many people have suggested that we use the actual brand names of products and list the hazards of the specific chemicals they contain. This would not be possible for several reasons. Many manufacturers consider their ingredients list to be proprietary information. That means they do not divulge the formulations. Another reason this would not be feasible is that manufacturers frequently change the formulations of products to save money, increase effectiveness, or improve the product in some way.

Through the years, toxicologists, those who study chemicals and their effects, have come to some agreement on doses and their results. The following table is based on consumption by a 150-lb person.

	Possible oral human lethal dose	
Toxicity rating	Dose	For 150-lb person
6: Super toxic	Less than 5 mg/kg	Less than 7 drops
5: Extremely toxic	5–50 mg/kg	Between 7 drops and 1 tsp
4: Very toxic	50–500 mg/kg	Between 1 tsp and 1 oz
3: Moderately toxic	0.5–5 gm/kg	Between 1 oz and 1 pt
2: Slightly toxic	5–15 gm/kg	Between 1 pt and 1 qt
1: Practically nontoxic	Above 15 gm/kg	More than 1 qt

After the publication of the first edition of *1001 Chemicals in Everyday Products,* we did a series of interviews on radio and television shows across the United States. Many of the shows had call-in sessions after the interview during which the listeners and viewers could ask questions. The first part of this book is based on those questions and answers and is the source, or inspiration, for much of the material in this second edition. Many of the frequently asked questions (FAQ) are answered in this book. As a result, we want to thank readers and radio and television listeners and viewers for the idea of using this material for the new section.

The second part of the book is an alphabetical listing of chemicals that compose everyday products. Most of the chemical entries include a listing of synonyms. Some of the synonyms include a CAS number. The CAS (Chemical Abstracts Service) identification number is being used more frequently on consumer products because it is the surest means of identification when using reference sources.

Readers frequently ask which is the greater danger: the risk of cancer from preservatives or the risk of sickness from bacteria that might be present in the food if preservatives are not used? There is no simple answer. Consumers must educate themselves and make decisions about the risks. We hope this book will help people make those choices.

CONSUMERS' CHEMICAL QUESTIONS

FOOD ADDITIVES

1. **What are chemicals used for in food?**
   Many times they are used to replace flavorings, colorings, and vitamins lost in processing.

2. **Why do so many chemicals have long complicated names?**
   The names are descriptions, or roadmaps, of the molecular formula.

3. **How is it that a chemical, such as hydrochloric acid, can be used in so many diverse ways? For instance, it is used for hair bleaching, swimming pool chemicals, and in food processing.**
   It depends on the concentration of the chemical. For instance, in food processing it is used in minute quantities to adjust the pH, the acidity, or alkalinity, of the product.

4. **Why is it some food packaging does not list what is inside?**
   There are some 300 standard foods (catsup, ice cream, and peanut butter, for example) that do not require listing of ingredients unless nonstandard colors or chemicals are added.

5. **If chemical additives are so dangerous, why are they permitted in food?**
   They are not dangerous for everyone. They do keep our food safer and cleaner. Because of allergies, sensitivities, and intolerances, some people must avoid them.

6. **How many food, cosmetic, and cleaning products are in the supermarket?**

Twenty years ago there were 8000; now there are estimated to be 24,000 different products.

7. **What are indirect food additives?**

These are chemicals, usually preservatives, that are put into food packaging so they can migrate to the food.

8. **What is the advantage to the manufacturer of putting preservatives in the food via packaging?**

The advantage is that the product has a longer shelf life and looks better (more attractive) to the consumer.

9. **What is the quantity of food additives we consume each year?**

It is estimated that we eat a total of 115 pounds of additives a year. This would include sugar and salt.

10. **I have heard that fruits and vegetables are fumigated in trucks and rail cars to kill insects and bacteria. Is there some alternative to this?**

One possibility would be food irradiation. It is used in 30 European countries and approved by the FDA and World Health Organization as safe.

11. **Would the food that will be hit with gamma rays to kill insects and bacteria be radioactive?**

No, it will not. The vitamins A, B, C, and E may be destroyed in the process though.

12. **Will irradiated food have labels that indicate it has been irradiated?**

Tomatoes, onions, strawberries, and mushrooms will need that irradiation label. In the past, spices, wheat, pork, and potatoes have not been identified as irradiated.

13. **While in England, I saw they used cyclamates. Why don't we use them in the U.S.?**

Cyclamates are being studied here and may be approved in the future. We use saccharin in limited amounts but England does not.

14. **What is another chemical that is used in the U.S. but not in England?**

Another example would be potassium bromate (a dough conditioner in bread, among other uses) which is not approved in England.

15. How many people are poisoned each year?

There were nearly two million cases reported to poison control centers last year. Most were accidental poisonings of children under five.

16. What was the major cause of poisonings of children under the age of five?

In most cases the children consumed too many vitamin or mineral pills. Iron pills look like candies and are very toxic in excessive amounts.

17. What are the symptoms of food allergy?

The most common is skin rash or hives. Other symptoms can include cramps, diarrhea, nausea, vomiting, sneezing, hay fever symptoms, and the most dangerous: anaphylactic shock.

18. Which foods are the most allergenic?

The most frequent offenders are peanuts, other nuts, soybeans, fish, shellfish, milk, and wheat.

19. Whenever I start at a new doctor's office they inquire about food allergies. Is this really necessary for them to know?

It is important for the doctor and staff to know about your allergies. For instance, recently it was determined that people who are allergic to fruits such as avocados, bananas, kiwi, papaya, or peaches are frequently the same patients who are allergic to latex. Of course latex is commonly used in medical offices.

20. How could fruit allergy trigger the same things that trigger latex allergy?

Latex is a natural product produced from the sap of the rubber tree in Brazil. It is thought that proteins in the sap are similar to those in the fruits.

21. What harmful effects could occur from a bad allergy attack?

The most serious allergy effect would be death from anaphylactic shock. Of course some people start with rashes and could have asthma attacks also.

22. Someone said that flounder, sole, and halibut are healthier to eat than tuna fish. Why is this so?

They could be referring to the fact that detectable levels of mercury were found in tuna fish, but not in the others mentioned, which are deep-water fish.

23. What are intermediate chemicals?

These are used in the processing of food. For instance, there are defoamers used in fruit juice. There are solvents used to remove caffeine from coffee, and antibacterials used to prevent bacterial growth in sugar.

24. What is PKU?

The letters stand for phenylketonuria. People with this condition cannot metabolize phenylalanine and must avoid the nonnutritive sweetener aspartame (Nutrasweet).

25. Can someone just have a simple allergy to aspartame?

Absolutely. Allergists say that everyone is allergic to something at one time or another.

26. In what foods are sulfites used? What is their purpose?

Some examples would be in wine and salad bars. They prevent bacterial growth and also prevent browning of fresh greens.

27. Can sulfites harm me?

There are people, particularly asthmatics, who must avoid sulfites. Most food or drink containing sulfites must be labeled. There have been instances where people who were asthmatics went into anaphylactic shock and died from consuming sulfites.

28. If I rinse off my fruits and vegetables, won't that get rid of the pesticide residue?

Not necessarily. Some fruits and vegetables (apples, cucumbers, peppers) are coated with wax and the pesticides, fertilizers, and fungicides are sealed in the wax.

29. I thought that the most dangerous pesticides to humans were banned by the U.S. Government for use in the U.S.?

That is true, but chemical companies are still manufacturing them and selling them to South and Central American countries. This produce then comes back to us in the winter, when 40% of the produce comes from foreign countries that do not have the restrictions we do.

30. **Is there any other way pesticides can get into food?**

    Yes, they can be in the fruits and vegetables because they are absorbed by plant roots and plant surfaces. This is called systemic absorption.

31. **What are some of the waxes used on fruits and vegetables?**

    They are beeswax, carnauba, candelilla, paraffin, and shellac, to name a few.

32. **Aren't some chemical preservatives being replaced by natural products?**

    An example would be potatoes. They have always been dusted with chemicals to prevent sprouting and rotting during storage. Recently, many producers started to use a product made from jasmine flowers. This is a safe, natural product, which is pleasantly aromatic.

33. **Is Kombucha tea hazardous to my health?**

    The CDC said that drinking four ounces daily might be safe for healthy persons. However, those with health problems or those who drink excessive amounts may risk illness. It is prepared by fermenting yeasts and bacteria together for two weeks with black tea and sugar. This mixture is called a "mushroom." Deaths associated with its consumption have been from a metabolic disorder characterized by elevated levels of lactic acid.

34. **What is the GRAS list?**

    It is a list of certain items developed in 1958. The letters stand for Generally Recognized As Safe. It is not a guarantee of absolute safety because some of the additives were later tested and removed from the list.

35. **What are some examples of items on the GRAS list?**

    Ordinary spices, salt, sugar, and phosphoric acid, (which gives soda pop its zip) are a few.

36. **I read that toxicologists analyzed some mother's breast milk and found PCBs. What are these?**

    PCB stands for polychlorinated biphenyl. This is found in electrical transformers and components. It is used for insulating and as a heat exchange fluid and hydraulic fluid. It is on the EPA Extremely Hazardous Substances List. It is a confirmed cancer-causing chemical that is toxic and can cause severe liver damage. It has also been linked to behavioral problems in children.

37. **Is MSG (monosodium glutamate) harmful to me if I eat food with it?**

     Some people have no ill effects whatsoever. Others have a variety of reactions including headaches, hallucinations, dyspnea (shortness of breath), nausea, vomiting, dermatitis, mood changes, and even IBS (irritable bowel syndrome). Some believe it does not have an allergic effect but rather a pharmacologic effect. That means that everyone would develop symptoms if enough of it were consumed. It could be compared to alcohol; that is, we all have different tolerances.

38. **Is MSG in any other foods besides Chinese foods?**

     Yes, It is used in meats, poultry, sauces, soups, pickles, condiments, bakery products, and even candies.

39. **How will we know which tomatoes, and other foods, are gene altered?**

     These foods will be labeled when they come to market.

40. **What improvement is there in a gene-altered vegetable?**

     A tomato would look better and have a longer shelf life.

41. **I have read about BST and BGH, the hormones that are given to cows. Would the hormones and antibiotics that are given to cows and other cattle get into our body when we drink the milk and eat the meat?**

     This is a major concern right now that is still being studied. There are conflicting scientific studies on the issue.

42. **What difference does it make if we get the antibiotics in our food via cattle?**

     The problem is that we may become antibiotic resistant and then antibiotic medication will have little or no effect on the infections we get.

43. **Will pesticides harm a person if they are not swallowed but simply get on the skin?**

     Most of the pesticides used in the U.S. are contact poisons. That means they are absorbed through the skin. That is why children and pets should not be on a recently pesticide-treated lawn.

44. **Don't these chemicals just pass through our bodies?**

     Some of the harmful chemicals do not. For instance, pesticides are stored in fat cells. In fact, the same is true in animals. If you eat meat, cut off the fat. Also lead is stored in the bones and can cause mental retardation, kidney and liver damage, retard growth, and have other harmful effects. Inhaled asbestos can cause cancer years later. These are just a few examples.

45. I heard that our Florida strawberries are rejected in Canada because of the fungicides on them. Why are they distributed here?

The fungicide you are referring to is called Captan® and the Canadian government rejects fruits with more than five parts per million. Our government permits up to 25 parts per million. The fact is that Captan® is permitted at this level because it is economically beneficial to the farmers.

46. If we are very careful and remove the skins of the vegetables then are we safe from consuming chemicals like pesticides, bactericides, fungicides, herbicides, and so on?

No, because some chemicals are put on the plants and absorbed into the plant by the leaves and roots and get into the fruit and vegetables this way.

47. What is the chemical that is put on plants that is similar to Lysol disinfectant?

That is called orthophenylphenol. It is a fungicide and the active ingredient in Lysol disinfectant.

48. Are some chemicals put on foods to make them last longer in the stores?

Yes, a common one is a herbicide sprayed on potatoes to keep them from sprouting. One popular preemergent is chlorprophame.

49. What are some of the problems associated with lead consumption by children?

Four million children consume lead each year by eating paint chips, chewing crayons manufactured in China, drinking or eating from foreign pottery, or drinking water from pipes with lead solder, to name a few sources. These children suffer hearing losses, learning disabilities, and lower IQs.

50. What is the chemical component in antacids and laxatives that we all can buy over-the-counter?

That is magnesium. It is generally safe for self-medication although poisoning is more common than once thought because it is frequently undetected by doctors and misdiagnosed.

51. What are the symptoms of magnesium poisoning?

The symptoms include irregular heartbeat, confusion, muscle weakness, nausea or vomiting, and possibly coma. This is diagnosed by routine blood tests that check the level of magnesium.

52. Who are at risk for magnesium poisoning?

The elderly and people with a history of GI tract disease who take multiple medications are the most at risk. Those who only take antacids and laxatives occasionally would not have any problem. Those who take them regularly should be sure their physician is aware of this.

53. Why did the pharmacist tell me not to take my antibiotic with orange juice?

Orange juice and the acids in fruit juices may decrease the effect of certain antibiotics such as penicillin and erythromycin. For this reason, pharmacists recommend drinking water instead.

54. Would it be a good idea to take antibiotics with a drink of milk?

If the antibiotic is tetracycline, milk and dairy products can reduce its effectiveness. For that reason, it is best to take tetracycline an hour before or two hours after milk or dairy products.

55. What does "caffeinism" mean?

"Caffeinism" refers to the increase of symptoms such as nervousness, agitation, headaches, insomnia, and irritability in a person taking birth control pills.

56. What is the chemical in honey that can help my memory?

Acetycholine is a brain chemical that allows brain cells to communicate better and improves memory. Studies at UVA have shown that honey contains a form of glucose that causes the release of this chemical. It was shown to improve memory for an hour in both students and Alzheimer's patients.

57. Why is it taking so long for more irradiated foods to appear on the supermarket shelves?

The chief reason is that companies are worried about the consumer's perception of irradiated foods. There have been many false reports about harmful effects.

58. What is meant by the term "designer foods"?

This term refers to hypernutritious foods or nutraceuticals. When it was determined that phytochemicals played a role in cancer prevention, many studies were conducted to determine which substances were most effective. The food industry supports nutrition research in cancer prevention and the possibility of supplementing foods with phytochemicals.

59. What are the phytochemicals?

These are chemicals that are contained in all plant foods. In fact, the word "phyto" means "plant" in Greek. Interestingly, they are not vitamins or minerals. Phytochemicals are not nutritious in the usual sense. Some types are isothiocyanates, polyphenols, flavonoids, monterpenes, and organosulfides. They are credited with the treatment and prevention of the four major causes of death: heart disease, cancer, high blood pressure, and diabetes.

60. What are the beneficial chemicals found in cruciferous vegetables?

They are isothiocyanates. These vegetables include broccoli, cabbage, and brussel sprouts. They have been shown to be chemoprotective against cancer of the breast, liver, esophagus, lung, intestine, and bladder.

61. Is it true that tea can be a cancer preventative?

Many epidemiological studies have shown that green and black tea phenolics inhibit esophageal cancer. It is thought that tea can be beneficial in other ways as well. Studies are currently being conducted.

62. What are the beneficial chemicals in citrus peels and cherries?

They are monoterpenes. They seem to act as cancer inhibitors in the colon, breast, liver, and pancreas.

63. If I buy fruit and produce and prepare it in the food processor for my baby, isn't this a lot healthier for my baby than jarred baby foods?

Not necessarily. The food you start with from the grocery may be older due to shipping and exposure to light and heat than the produce the manufacturer gets from the field and immediately processes.

64. What do they add to the meat in fat-free hot dogs to keep them juicy?

They add water, gums, and colloids (starch-type products) to increase tenderness and retain moisture.

65. What is it about foods that contain fat that make them taste better than nonfat foods?

It is hard to believe, but fat molecules are round, and this is the sensation that our taste buds find so appealing. Protein molecules are elongated. We enjoy the smooth feeling in our mouths, according to food scientists.

66. Scientists say that women planning on having children and adults who are at high risk of heart attack, should take folic acid supplements. What are the recommended doses, and which foods are highest in folic acid?

The CDC suggests 400 mcg daily for those two groups of people. The highest mcg is found in lentils at 179 mcg per half cup and fortified breakfast cereals, which have 146 to 177 mcg per half cup. Spinach has 131 mcg per half cup, orange juice 109 mcg per cup, artichokes 95 mcg per cup. Green peas, broccoli, corn, and so on are all beneficial. Usually, natural sources are recommended. However, folic acid studies show that supplements are better because they are more easily absorbed the food-based folate acid.

67. What causes intestinal gas?

The major cause of intestinal gas is undigested carbohydrates. In particular, the oligosaccharides, which are the most common of the carbohydrates. These are mostly found in beans and other legumes. The second most common is lactose or milk sugar. The problem exists because so many people are lactose intolerant. This means they lack the enzyme that breaks down lactose.

68. What is the red pigment in tomatoes?

It is lycopene, which is a powerful antioxidant. Harvard and U.K. researchers' studies indicate that lycopene is thought to block cell damage leading to certain cancers. The studies are particularly favorable regarding prostate cancer. Therefore, cooked or fresh tomatoes may be as important in preventing cancers as carrots and green vegetables. The tomatoes contain significant amounts of beta carotene which is also considered to be an excellent antioxidant.

69. What are some of the other chemicals in tomatoes?

Tomatoes also contain phenolic acids and plant sterols. These are also considered to be protectors against cancer. They also contain monoterpenes. These are antioxidants that protect against cancer and prevent cholesterol production.

70. Is there any way to avoid the danger of salmonella when eating raw eggs or foods that contain raw eggs like Caesar salad?

Food scientists at Purdue University have figured out a way to pasteurize eggs in their shells. Currently, the only safe ones would be the liquid ones that have been removed from their shells and pasteurized.

71. Are there any natural food preservatives or are they always in the form of added chemicals?

There are naturally occurring food preservatives. One has recently been identified. It is a substance in cinnamon named benzaldehyde. It is also found in black pepper, vanilla, and peaches. We may be seeing it in the future as a preservative in canned foods that now require heating for lengthy times before they are safe.

72. What is chromium?

It is a trace mineral that is considered important in helping cells break down sugar into energy for the body.

73. I have read some wonderful reports about people taking chromium supplements. They said overweight people lose weight, gain muscle mass, live longer, and improve well being. Have these results been proved?

No, they have not. The scientific studies have produced mixed results. Toxicologists believe the metal may accumulate in tissues. Scientists suggest that a lower supplement, such as a 50-mcg dose, may be more appropriate than massive doses. This is sold as a supplement and does not require FDA approval.

74. What foods are high sources of chromium?

Cereals are good, especially bran cereals, as well as canned tomatoes and canned pineapple, broccoli, black pepper, tea, cocoa, corn chips, and white potatoes.

75. There is a new medicine for osteoporosis. Does this mean I can stop taking calcium if I take the new medication?

The chemical name is alendronate sodium. The prescription drug name is Fosamax®. Studies showed that it could actually restore bone mass. It normalizes the bone-building cycle, but you still need to take calcium to mineralize the bone.

76. Is the new medication better than the nose spray medication for osteoporosis?

Yes, so far it appears that it is an improvement over the other two treatments, calcitonin nasal spray and fluoride.

77. What are the hormone supplements that are so popular these days?

You are probably referring to chemicals such as melatonin and DHEA.

78. What is tryptophan?

It is an essential amino acid found in milk and dairy products. There have been claims that, since it increases the neurotransmitter serotonin, it is a relaxant and induces calmness. Studies have shown that oral ingestion or injection of tryptophan raises brain tryptophan and serotonin levels. However, foods rich in tryptophan, such as dairy products, do not increase brain levels or stimulate serotonin synthesis.

79. Is melatonin used in Europe?

Melatonin is banned in France and Britain. Other countries consider it a drug and it is only available via prescription. Countries such as Italy and Denmark have prohibited manufacturers from making claims as to its benefits until proved. In Europe these supplements are considered black market products. However, many Europeans acquire them from American friends. Even though they can be found in health food stores in the U.S., they are under-the-counter products in Canadian stores.

80. What does DHEA stand for?

It stands for dehydroepiandrosterone, which is called the mother hormone. It has been a "hot" supplement for the last couple of years.

81. What is DHEA?

DHEA is dehydroepiandrosterone. It is a plant derivative of a hormone from the Mexican wild yam. It is naturally produced by the adrenals and is one of the most abundant hormones in humans. It begins at age 7 and peaks around 30 years of age. As a supplement, it promises to enhance metabolic functions, increase energy, control stress, maintain proper mineral levels, balance the production of hormones, fight aging, improve memory, build body muscle, and reduce fat. It supposedly reduces the risk of cancer, heart disease, atherosclerosis, Alzheimer's disease, and schizophrenia. Many scientists claim this supplement is useless.

82. What are the chemical components of chocolate?

There are many components in chocolate. The biologically active stimulants are caffeine and theobromine. The amine compounds are tyramine and phenylethylamine or PEA.

83. What is the effect on the body of the amine compounds found in chocolate?

Amine compounds act as stimulants and could elevate blood pressure. They could also cause migraine headaches.

84. How does phenylethylamine work in our body?

It dilates our brain's blood vessels and triggers headaches. Red wines and cheese are also in this category and cause the same effects.

85. Can the chemicals in chocolate cause any other problems?

People who are at risk for kidney stones should not eat too much chocolate because it can increase urinary oxalate excretion.

86. I have heard that garlic contains many useful chemicals. What are they?

Garlic is a very special vegetable. It contains vitamins A, B, and C. It also has calcium, potassium, and iron. It contains antioxidants, carotenes, germanium, and selenium. Most important is the amino acid allicin. Garlic contains chemicals that act like ACE, that is, angiotensin converting enzyme inhibitors. These are the prescriptions usually given to lower blood pressure and protect the heart. It is thought that garlic works by dilating blood vessels.

87. How do these chemicals in garlic help our bodies?

These ingredients have been credited with having antibiotic effects, helping asthma patients, and combatting acne problems, ear infections, gallbladder disorders, and diarrhea. They also are said to cause reductions in HDL and even be inhibitors of fibrogen that causes platelet clumping that leads to stroke and hardening of the arteries. Garlic was used by the British during World Wars I and II to prevent gangrene and treat infections.

88. Is it true that chemicals in dark beer make it healthier for a person to drink than light beer?

Dark beers are very high in flavinoids. These contain some vitamin-like compounds that researchers claim make platelets less likely to clot.

89. What is the story on Olestra®?

Olestra® is a fat replacer food additive. It is a sucrose polyester that can be used in place of fat without adding calories. As a result of its chemical composition, it is not digested by the body and passes through without adding fat or calories to food. That means that it is not digestible.

90. Is Olestra® approved by the FDA?

Yes, it has been approved, but it requires a caveat also. An Olestra® product must include the following statement on the label: "This product contains olestra. Olestra® may cause abdominal cramping and loose stools. Olestra® also inhibits the absorption of some vitamins and other nutrients. Vitamins A, D, E, and K have been added." FDA only approves its use in snack food.

91. What is meant by functional foods?

This term refers to foods that contain physiologically active components that promote health and may prevent disease. Most foods are functional because they contain chemical nutrients that are essential for health. Currently there is a great deal of research on the phytochemical components of foods such as beta carotene and other cartenoids that are responsible for good health.

92. Is it true that a poison such as sulfur dioxide is used on grapes?

You are confusing the colorless gas sulfur dioxide with the chemical that is used on foods as a bleaching agent or preservative. It is mildly toxic to humans when inhaled as a gas, but after being used on grapes to prevent darkening, it dissipates and is harmless to the consumer. This is why we have golden raisins; when sulfur dioxide is used, grapes do not darken as regular raisins do. They are also dried with artificial heat to prevent darkening.

93. Isn't produce sometimes exposed to ethylene gas for a purpose?

Yes, many times tomatoes are picked green and ripened with ethylene gas. These are the flavorless ones as opposed to vine-ripened tomatoes.

94. Is it the alcohol or other chemicals in wines that seem to prove beneficial to the heart in moderate amounts?

Some chemists are saying that the phenolics, natural antioxidants, in nonalcoholic wine prevent cholesterol molecules from forming and clogging the arteries. Phenolics are also found in fresh fruits and vegetables such as raisins, grapes, and onions. At the same time, other scientists say benefits are derived by beer drinkers; therefore, it must be alcohol that provides the benefits.

95. I am pregnant and a heavy coffee drinker. I recently heard that another bad effect was attributed to caffeine. What is that about?

Recent studies indicate that pregnant women who drink three or more cups of coffee or tea a day during the first trimester of pregnancy have twice as many miscarriages as those who drink less.

96. How much caffeine do the popular beverages contain?

Here are some figures on that subject from the FDA and National Soft Drink Association. A 5-oz cup of coffee averages 115 mg of caffeine, a 5-oz cup of tea 40 mg, a 6-oz soft drink 18 mg, a cup of cocoa 4 mg, and one ounce of chocolate syrup 4 mg.

97. I heard that tea contains a lot of chemicals. What are these?

More cultures of the world drink tea than any other beverage. The tea that is most popular and with which we are most familiar is obtained from the leaves of an evergreen tree. Other teas are produced from herbs, flowers, and roots. Some people have allergies to these teas. For instance, chamomile is a plant cousin of ragweed and could produce an allergic reaction. Tea leaves contain fluoride, which is beneficial to our teeth. Polyphenols, chemicals considered to be anticancer and heart disease, are also abundant. The ingredient tannin can be irritating and can cut down the absorption of iron although moderate amounts are probably harmless. The flavonoids in tea are thought to prevent strokes.

98. Why is salt iodized?

We use iodized salt to prevent goiters and mental retardation. In China, a country of 1.2 billion people, there are 10 million cases of retardation from lack of iodine. Although just trace amounts would prevent this, many people buy noniodized salt on the black market to save money. Maternal iodine deficiency results in growth and brain impairment in the child.

99. I know that sodium is a chemical element. What does it mean when a label says the food has low sodium?

This term refers to food that has no more than 140 milligrams of sodium per serving. Therefore, it is more desirable for someone who is advised to be on a low sodium diet.

100. When we see "salt free" on the label, is this the same as low sodium?

If it says "salt free," the food contains fewer than five milligrams of sodium per serving.

101. What does it mean if a label states that the food or drink is sugar or sucrose free?

This is defined as less than half a gram of sugar per serving.

102. **Can we be sure that a food that says it is calorie free has absolutely no calories in it?**

No, by definition, "calorie free" refers to a food with no more than five calories per serving.

103. **Since low-fat diets are so popular, what does it mean when the label says "fat free"?**

"Fat free" on a label means less than half a gram of fat per serving, assuming the food has no added fat or oil ingredients.

104. **Labels sometime state that the product is low fat. How much fat is really in it?**

This would mean that it contains no more than three grams of fat per serving.

105. **Another common term that is often seen on labels is the word "light." What about these foods?**

They contain at least one-third fewer calories, or half the amount of fat, in the usual or typical version of the food.

106. **How can the "questionable" chemicals be sold right over the counter in health food stores?**

There is a law that states that if a product is labeled as a "food supplement" the product can bypass some restrictions required for drugs and chemicals.

107. **Why is BHA added to meats, cereals, chewing gum, desserts, shortening, dry fruit, margarine, pizza toppings, potato products, poultry, rice, and sausage?**

It is added to foods to act as a preservative and antioxidant, to slow down spoiling, and to extend their shelf life. Unfortunately, it has caused cancers, tumors, and birth defects in animal studies. Its use is limited by FDA regulations.

108. **What are the major components of foods?**

Water is the major component. In fact, 99% of foods are composed of water, proteins, carbohydrates, fiber, fat, and ash.

109. **What are the minor components of foods?**

Vitamins and minerals are the minor or microcomponents.

110. **Should we be concerned about the bacteria in food?**

Food plant sanitation has greatly improved over the years. Some germs are controlled in foods by heat processing, freezing, dehydration, chilling, and ionizing radiation. Remember that not all bacteria are dangerous. Bacteria are used to ferment foods. That is how some pickles, sausages, alcoholic drinks, and yogurt are prepared.

111. Which food has the greatest water content, celery, tomatoes, or milk?

Surprisingly, celery has over 94% water content, tomatoes have over 93%, and milk has just 87%. The water in the vegetables is held in the cell walls.

112. Why is the FDA requiring manufacturers to fortify products with folic acid?

This is to prevent spinal cord birth defects. Women of child-bearing age should consume 400 micrograms a day. It is also beneficial to others to cut the risk of heart disease and cervical cancer. Since supplemented foods will still not supply enough, it is recommended that individuals also consume orange juice, dried beans, and wheat germ.

113. What is DHA or omega-3 oil?

This is docosahexaenoic acid, a brain protective fatty acid derived from fish. It also is the same oil that has been credited with fighting heart disease. Some of the beneficial results are reducing depression and aggression, and as a stimulant to brain development.

114. In what foods can I find vitamin B6 and folic acid?

Vitamin B6 is found in whole grains, beans, legumes, fish, pork, and chicken. Leafy vegetables, orange juice, legumes, and fortified cereals are the best sources for folic acid.

115. What is this new product in the health food stores called pregnenolone?

This is promoted as nature's feel good hormone. It is a hormone made from cholesterol. It is formed in the body organs that produce steroid hormones, such as the liver, adrenals, skin, ovaries, and testicles. It is thought that the brain has the capacity to use cholesterol to make pregnenolone and other steroid hormones. Some believe this chemical has a positive effect on energy, memory, stress reduction, and arthritis. It is thought to be beneficial as a hormone replacement for those in middle or old age.

116. What is the chemical that gives onions such a strong smell?

They contain substances called thiols that are characterized by strong odors. Of course onions are edible by humans, but dangerous to dogs and cats.

117. **What are the harmful chemicals that are formed from grilling meats?**

    PAHs or polycyclic aromatic hydrocarbons are formed when fat drips down into an open flame that sends up smoke, coating the foods with carcinogens. HCAs are created when meat, fish, and poultry are cooked at high heat such as when well done, panfried, barbecued, or broiled. There is a chemical in muscle meats called creatine that produces the HCAs. The safest way to prepare such foods would be by microwaving, stewing, poaching, or boiling.

118. **What is a nutraceutical?**

    This is the term given to foods that are rich in antioxidants and phytochemicals. These are foods that provide benefits beyond the traditional vitamin and mineral content.

119. **What is the current thinking about calcium and kidney stones?**

    The newest studies indicate that calcium in milk may be beneficial. It is thought that calcium helps to neutralize the effect of oxalates, the chemicals that are thought to initiate the formation of kidney stones in some people. Oxalates are in fruits, vegetables, whole grains, and beans. When the foods that contain oxalates are consumed with milk, the calcium in the milk binds with the oxalates, and allows the food to go through the body without absorbing the oxalates. In this manner the risk of kidney stones is reduced.

120. **What are the benefits of soy isoflavones?**

    Soy isoflavones are a group of phytochemicals only found in soybeans. The isoflavones genistein and daidzein have strong antioxidant benefits that reduce the risk of heart disease and cancer, as well as other diseases.

121. **Do soybeans contain other beneficial chemicals?**

    Soybeans contain saponins that are a group of phytonutrients with immunity and disease-fighting potential. They also act to increase immunity, prevent cancer, and fight infections. Soy is also high in protein, vitamin E, lecithin, and omega-3 fatty acids.

122. **What is the newly identified nutraceutical in cereals?**

    Tocotrienals are the new antioxidants found in the oil fractions of cereal grains. Examples would be wheat, rice, rye, and barley grains. The benefits of tocotrienals are similar to those of tocopherols and vitamin E. They can reduce cholesterol and slow the advance of atherosclerosis and the progress of certain cancers. These nutraceuticals are available as supplements.

123. What are the flavonoids in wild blueberries?

They are anthocyanins. These flavonoids are the blue pigment material in the fruit. This is an antioxidant that has been credited with improving eyesight, slowing the effects of aging and memory loss, and improving circulation.

124. What is Pycnogenol®?

This is a supplement containing more than forty water-soluble antioxidants. It is a complex of antioxidant flavonoids obtained from the bark of French maritime pine trees. These flavonoids are credited with reducing stroke and heart disease and improving the immune system to effectively fight viruses.

COSMETICS

1. What does the FDA state that a cosmetic can claim?

A cosmetic can claim four things. It can cleanse, beautify, promote attractiveness, and alter the appearance in general.

2. What type of claims can be made for drugs?

A product that makes structural and functional claims can be classified as a drug. For instance, a product that penetrates the skin and causes cell rejuvenation has legitimate claims as a drug.

3. How are vitamins categorized, as drugs or cosmetics?

According to the Dietary Supplement Health and Education Act of 1994, this has not been defined. These are probably dietary supplements that the FDA no longer oversees.

4. When does a soap become a cosmetic?

If the manufacturer makes a cosmetic claim on the label, such as "moisturizing" or "deodorizing", the product must meet all FDA requirements for a cosmetic and the label must list all ingredients.

5. When does soap become a drug?

If the manufacturer makes a claim that the soap has antidandruff, antibacterial, antiperspirant, or antiacne action, the product is a drug, and the label must list all active ingredients as is required for all drug products.

6 **Are the ingredients in cosmetics and personal care products approved by the government?**

There is an industry sponsored Cosmetic Ingredient Review (CIR) that tests the ingredients. Essentially, this is a self-regulated industry.

7. **Are the manufacturers of cosmetics and personal care products required to list the ingredients?**

Yes, hair sprays, creams, shaving lotions, and perfumes should list ingredients. On the other hand, this applies to items manufactured in this country. If they are imported, they might not have an ingredient label.

8. **What ingredient in cosmetics causes the most dermatological problems?**

A mascara preservative called quaternium 15 is one of the biggest offenders and cause of contact dermatitis. Of course, the eye area is the most delicate.

9. **The ingredient, methylparaben, seems to be in many cosmetic products. What does it do?**

It is a preservative that prevents mold growth. It is frequently a cause of allergic reactions.

10. **Why is the list of ingredients so long on a simple bottle of shampoo?**

The manufacturers want to add coloring, perfume, preservatives, sequestrants (holds it together), fillers, humectants (moisturizers), thickeners, texturizers, and stabilizers. These make the final product appealing to the consumer.

11. **If the cosmetic says it is hypoallergenic is it completely safe to use?**

That means there are no fragrance ingredients in the product, but someone could still be allergic to other ingredients.

12. **Why can I not use hair dye on my eyelashes and eyebrows?**

Coal tar dyes, which are legal for hair dyes, are restricted from use around the eyes. They have caused blindness.

13. **Are alcohol-based mouthwashes safe to use?**

One study indicated that high alcohol content products can increase a person's risk of oral cancer by as much as 60%. More recent studies have questioned this.

14. **What is the harmful ingredient in nail polish?**

It is toluene. It is mildly toxic by breathing and can cause hallucinations, CNS (central nervous system) effects, bone marrow changes, and birth defects, among other problems.

15. What other products contain toluene?

Toluene is also in paints, spot removers, rubber cement, gasoline, detergents, and perfumes, among others.

16. What is meant by aroma-free zones?

Some feel that the odor of cologne, perfume, and shaving lotions are offensive and dangerous to those with allergies. There is talk that there will be areas of restaurants and public places set aside for those who do not want to be exposed to strong aromas.

17. Is the aluminum chloride in deodorants harmful to us?

It is a possible allergen and irritant. It would be harmful if it were swallowed. There is still some suspicion that aluminum could be a factor in Alzheimer's disease, but it has not been confirmed.

18. What is the greatest problem from the use of cosmetics and toiletries?

Probably the absorption of solvents, coloring agents, and fragrances. It once was thought that the skin was a barrier that did not absorb chemicals. We now know more about cutaneous absorption. Of course, allergies are a problem, too.

19. Why is talcum powder dangerous?

It could be an allergen and skin irritant. If inhaled repeatedly, it could cause pulmonary fibrosis that may be due to asbestos in the powder. Also, a study of women who used talcum powder on their sanitary napkins indicated that they had an increased incidence of ovarian cancer.

20. What is being substituted for the chlorofluorcarbons (CFCs) as propellants in air fresheners, deodorants, hair sprays, and so forth?

Propane and butane are two commonly used propellants. These are also harmful because when inhaled they depress the central nervous system and may lead to cardiac or pulmonary arrest. Permanent nerve and brain damage could occur.

21. For what is formaldehyde used?

It is a deodorizer, solvent, disinfectant, and adhesive that is in glues, air fresheners, antiperspirants, dry-cleaning solvents, hair spray, after-shave lotions, permanent press fabrics, particle board, and plywood.

22. Is formaldehyde harmful?

Yes, it is thought to be a cause of sick building syndrome. The air inside buildings and mobile homes can be dangerous because of the outgassing of chemicals such as formaldehyde from the paneling, carpeting, and wallboard. Long-term exposure affects the central nervous system. It could cause aggression or drowsiness. It can cause skin and eye irritation. It is a confirmed cancer-causing chemical. If swallowed, it is a poison that causes violent vomiting and diarrhea.

23. What are some harmful effects to cosmetologists from using products containing formaldehyde?

A recent study indicates that they have twice as many miscarriages as other women, from using the formaldehyde-based products in their profession, such as hair perms, nail products, and disinfectants.

24. How does the new vitamin C cream help to maintain youthful skin?

This is the newest entry after Retin A and alpha hydroxy acids (AHA). It is a nonprescription, reformulated vitamin C serum that can penetrate the skin to help repair collagen, which is the network of fibers supporting the skin. It also protects the skin from brown sunspots.

25. How does sunscreen work?

Some sunscreens are actual blocks to the sunlight. For instance, titanium dioxide will act as a block and will prevent the sun's rays from reaching the skin. Other chemicals act as UV (ultraviolet) absorbers, thus preventing the harmful effects of the sunlight.

26. Is nail polish harmful to the nails?

Our fingernails, like our hair, are composed of dead cells. In most cases applying chemicals to these is not harmful to the body. Sometimes chemicals in nail polish, such as formaldehyde, can affect the surrounding skin. Also chemicals applied to the hair, such as calcium thioglycollate, can affect the skin of the scalp.

27. How do we get sufficient vitamin D without exposing our skin to harmful sun rays?

Dermatologists all agree that we should avoid the sun. However, it is beneficial to expose our face, arms, and hands to sunlight for only 10 to 15 minutes two or three times during the week. Vitamin D is stored over the winter in our bodies. Multiple vitamins should contain 10 mcg or 400 IU of vitamin D.

28. What are some of the chemicals recommended by physicians for skin rejuvenation?

The first one that was recognized as being effective was retinoic acid. This was a prescription ointment drug until the over-the-counter product was approved in the mid-nineties, and proved effective. Another is glycolic acid, which is available in makeup products or in more concentrated forms from physicians. There is also Kojic acid, alpha hydroxy acid, beta hydroxy acid, and the newest vitamin C products. Hydroquinone is another chemical that is used because it has a bleaching effect on the skin.

29. As a swimmer, I find that my dyed hair fades a lot. What chemical causes this?

If you swim in the ocean, it is the salt; if you swim in a pool, it could be the chlorine. Also, the cheaper high-alkaline shampoos cause hair drying.

30. What is in those lotions that are supposed to make the skin look suntanned?

The active ingredient is a chemical called dihydroxyacetone. It works by reacting with the protein in the skin, which causes the skin to have a tan or brown appearance.

CLEANING PRODUCTS

1. Do household cleaning items require labeling to tell us what is in the product?

No, labeling is required only if it is flammable, corrosive, or toxic. It must then be labeled "dangerous" or "poison."

2. Do air fresheners really freshen the air?

No, they cover up the offensive odor with a stronger odor. The sprays are oil based and when inhaled, the oily substance covers the olfactory and smelling surfaces and prevents us from smelling the other odors.

3. Should we launder new clothes before we wear them?

Some people should. Most clothing and fabrics are coated with a chemical finishing agent that some people find irritating or the cause of allergic reactions. This is particularly true with permanent press.

4. What is the most common problem encountered when using cleaning products?

The most common problem is the result of people using products without adequate ventilation.

5. What is the chemical used by dry cleaners?

The most common one is "perc," or perchloroethylene. It is used in 85% of the 30,000 dry-cleaning shops. They emit 92,000 tons of this into the air each year. It is also used in metal polishes and degreasers.

6. Is this perchloroethylene chemical harmful?

It is a probable carcinogen (causes cancer) on the EPA list. When breathed, it is moderately toxic, causes eye and skin irritation, and possibly hallucinations, coma, and lung changes. People who live near dry cleaners have been exposed to "perc" levels hundreds of times higher than the acceptable guidelines. Newly dry-cleaned clothing should be hung out in the open for airing, and never hung in children's rooms. Children are more sensitive to toxic substance.

7. What is the most dangerous chemical around the house?

The most dangerous one is the one in an unmarked container. Many times people transfer the contents from the original container to another bottle or jar that is not labeled. The tragedy of children consuming or using something like this is well known by the poison control centers.

8. Is it dangerous to mix cleaning products?

Yes, that should be avoided. For instance the common mistake of mixing plain household bleach and ordinary ammonia to clean a sink or bathtub can produce a deadly phosgene gas.

9. What are some other strong cleaning chemicals we should know about?

The caustics such as sodium hydroxide in oven cleaners or drain cleaners are very toxic and corrosive. This kind of product will have definite instructions and warning labels that should be followed carefully.

10. What are some other names for caustics such as sodium hydroxide?

They are caustic soda, lye, soda lye, sodium hydrate, sodium hydroxide, and white caustic.

11. Isn't there a lot less pollution in the air we breathe since the automobiles have all those catalytic converters and air pollution control devices?

Yes, at least the levels of CO_2, or carbon monoxide have been reduced from motor vehicles. Some scientists say that it would be hard to asphyxiate oneself with a new, well-tuned vehicle.

12. Are some states better than others when it comes to labeling laws?

Yes, both California and New Jersey have stricter laws than the other states.

13. How did the labeling laws make products safe?

Obviously it is better for people to be informed, but also some products were actually reformulated after the labeling laws were passed so consumers would not complain about the harmful ingredients.

14. What is an example of a product that was reformulated after the labeling law was passed?

Trichoroethylene, a cancer-causing chemical, was removed from correction fluid. Percloroethylene was removed from spot remover.

15. Just because a chemical causes cancer in laboratory mice does that mean we should avoid it?

Materials that cause cancer in one type of animal usually cause cancer in other animals according to the Cancer Institute. To prevent cancer we cannot wait for absolute proof of carcinogenicity in humans.

16. What does it mean when it is said "The dose makes the poison"?

This refers to the fact that some chemicals are harmless in the specific amounts necessary for a particular purpose, but they are harmful when used or consumed excessively.

17. Since we are indoors most of the time, are we exposed to more chemicals?

Yes, in fact the EPA thinks that indoor pollution is one of the major health risks of the decade. We have had several episodes of the sick building syndrome already.

18. **How can we avoid the problems of inhaling the outgassing of formaldehyde from wood paneling, fumes coming from freshly dry-cleaned clothing, and the other indoor pollutants?**

   The old saying is that the "solution to pollution is dilution." This means that it is good to open windows or use ventilation fans.

19. **What is the ingredient in toothpaste that causes canker sores in some people?**

   SLS or sodium lauryl sulfate is an emulsifier. An emulsifier is something added to a formulation to prevent ingredients from separating and decomposing. Sodium lauryl sulfate (SLS) is a common ingredient in toothpaste, shampoos, detergents, and so on. A recent study indicated that it may dry out the mucous membrane of the mouth, causing canker sores in susceptible individuals.

Miscellaneous questions

1. **What is the single most important chemical?**

   This is really difficult to answer because chemicals do so many different things. In one respect, chlorine is the chemical that has saved more human lives than any other as a result of its water purifying effects.

2. **If I have a choice of paper towels, hot air dryers, or pull-down linen towels to dry my hands, which should I use to avoid bacteria?**

   There have been English studies that showed that most hot air dryers blew out bacteria that could cause bronchial pneumonia. Many of them blew out bacteria that indicated fecal contamination. Linen towels reduced the bacteria on people's hands by more than 40% and paper hand towels by nearly 60%.

3. **What is melatonin?**

   Melatonin is a hormone that is secreted by the pineal gland, a pea-sized organ in the center of the brain. Originally it was used in low doses to hasten sleep and overcome jet lag. More recently, people have begun thinking of it as a fountain of youth.

4. **Has the government approved melatonin?**

   Melatonin is sold as an over-the-counter supplement. As a result it is a dietary supplement and not a drug, so it is not subject to testing and approval by the Food and Drug Administration.

5. Is it known whether melatonin is definitely harmful to some people?

There are many who should not take it; these include children, nursing mothers, pregnant women, those with conception problems, mentally unstable people, those taking steroids, cancer patients, people with severe allergies, and those with autoimmune diseases. This is the reason a person must check with a physician before taking supplements.

6. From where is melatonin derived?

It is derived from an amino acid called tryptophan. Melatonin does influence serotonin and dopamine, neurochemicals in the brain.

7. What are amino acids?

These are the building blocks of protein. There are twenty considered necessary for our health, and eight of these must be obtained through the diet.

8. What is astragalus?

This is a plant that grows in the western U.S. It is also called locoweed and milk vetch. Recent research has suggested that it shows promise as an immune booster for cancer patients. It is toxic to sheep, cattle, and horses when ingested.

9. What is echinacea?

This is an extract of the purple coneflower plant. It is a popular item in health food stores for self medicating. European studies indicate that it reduces the severity of cold and flu symptoms and is believed to increase the production of white blood cells. It is used by Native Americans as an antiviral agent. It should not be used for more than seven days as it can suppress the immune function.

10. What is ma huang?

This is a controversial central nervous system stimulant. The source is the dried leaves of a Chinese plant. It is used in sports nutrition and diet products. It contains ephidrene, an alkaloid found in decongestants and asthma medicine. Death has resulted for some people who took it and had heart, thyroid, circulatory problems, or, in some instances, combined it with caffeine. In many states pharmacists only sell it to people 18 or older.

11. **Are the herbal mixtures sold in magazines and at concerts to teen-agers as a natural alternative to street drugs dangerous?**

They definitely could be. The products usually contain stimulants such as ma huang, ephedra, kola nut, green tea, and guarana. These all contain caffeine. These combinations have caused heart attacks, strokes, and nerve damage as well as death in some individuals.

12. **What is chitin?**

It is a substance found naturally in the shells of crustaceans, such as crab, shrimp, and lobster. It is also found in the exoskeleton of marine zooplankton, such as coral and jellyfish. Insects also have chitin in their wings. Examples would be butterflies and ladybugs. The cell walls of mushroom and other fungi contain this also. This substance is used in food, cosmetics, and biomedicine. It has been found to be an antibacterial, antifungal, and antiviral and used for wound dressings, sutures, cataract surgery, and periodontal disease treatment. Its many sources and uses make it a fascinating material.

13. **What are some of the beneficial insects that replace chemical insecticides in the garden?**

Examples of beneficial insects are syrphid flies, lacewings, parasitic wasps, and ladybugs, which eat the mites that attack plants.

14. **What is diatomaceous earth and what are its uses?**

It's a powdery material formed from the shells of one-celled sea plants such as diatoms. Each particle is sharp and can kill insects by abrasion. This is its mode of operation as a pesticide.

15. **What is the name of the metal that can burn?**

Magnesium is the metal that can burn and the dust can cause serious explosions, as all dust can. When the fires become large in an industrial situation they cannot be extinguished except by smothering, that is, by excluding oxygen from the site.

16. **Why are the chemicals in some antibiotics not as effective against diseases as they once were?**

Some people take antibiotics for viral sicknesses, such as head colds, without determining that the cause is bacterial. Antibiotics will not cure viral sicknesses and our body becomes resistant to their effectiveness when we really need them. Some people stop taking medication too soon and the antibiotics do not completely kill off all the bacteria and mutate into a new strain. Antibiotics should only be prescribed for bacterial infections and the patient should take all of the medication.

17. Some athletes are taking a chemical called vanadyl sulfate to produce the bulked up look attained via steroids. Is this safe?

Vanadyl sulfate is a cloth or ceramic dye that is not meant for human consumption. There is a product being sold to athletes, who believe it is a wonder drug. The fact is that this chemical can cause the pancreas to reduce the production of insulin and develop a type II diabetic state. Until further studies of this product are released, it would be advisable to avoid it.

18. What are the disease-fighting chemical nutrients referred to as flavonoids and retinoids?

If you eat green salads, you will be consuming these disease-fighting nutrients. They are also found in blueberries and grapes in high concentrations. Flavonoids are compounds found in plants. They have antioxidant and antiinflammatory effects.

19. We bought some bags that are supposed to keep our fruits and vegetables from rotting as fast. From what chemical is this bag made and how does it work?

These green plastic-like bags are made from oya, a natural mineral that absorbs ethylene gas. The gas causes the rotting.

20. What is the chemical in those little birthday cake candles that reignite after you blow them out?

They have magnesium crystals in them, which retain the heat even after the flame is out. The heat is great enough to reignite them. Obviously this is a dangerous fire hazard if handled improperly.

21. What is the chemical used as a deterrent for the problem of excessive wild Canadian geese in some areas?

Methyl anthranilate, a chemical that occurs naturally in orange blossoms, jasmine flowers, and grapes. This substance is used in making saccharin and chewing gum flavoring. However, geese are repelled by grass sprayed with this substance and move on to other ponds and grassy areas. This prevents excessive bird droppings in suburban parks. It is otherwise harmless to the birds.

22. What chemicals can be added to cut flowers to keep them fresher longer?

Take a pint of warm water and one pint of lemonade or lemon-lime soft drink, mix them together, and be sure the bottom three or four inches of the stems are covered. This works because the sugar in the mixture supplies nutrients to the flowers, and the citric acid acts as a preservative.

23. If we consume toxic chemicals we can get sick or even die. Are there some plants that would harm my pets in the same way?

Yes, we have a list of common plants that are toxic to pets. This list does not include all the harmful plants, just the most common ones. They are amaryllis, arrowhead vine, asparagus fern, azalea, bird of paradise, Boston ivy, chrysanthemum, creeping Charlie, creeping figs or ficus, crown of thorns, dieffenbachia or dumb cane, elephant ears, emerald duke, ivies, Jerusalem cherry, pothos, philodendrons, pot mums, red princess, spider mum, sperengeri fern, umbrella plant, weeping fig, mistletoe berries, poinsettia; also most plant bulbs.

24. What is the chemical used in the manufacture of toys to make them antibacterial?

That is Microban®, which permanently bonds tiny germ-killing pellets to plastic or fiber. It is used in highchair trays and plastic toys. Actually, hospitals have been using products with Microban® such as surgical drapes, mattresses, and pillowcases, for over ten years. The newest products include food-cutting boards, athletic shoes, and carpeting that will have Microban® fighting microbes. Microban® stops mold, mildew, fungi, and bacteria, including E coli, staph, salmonella, and strep.

25. What is meant by homocysteine levels?

This is an amino acid that is a result of high protein diets. It is thought to be as harmful as high cholesterol. A level above 14 is considered dangerous. This can be reduced by consuming folic acid supplements.

26. What is the best way to break down the harmful amino acid homocysteine in the blood?

The latest heart healthy information advises us to eat a variety of foods rich in vitamin B6 and folic acid. These vitamins help to break down homocysteine.

27. What is coenzyme Q10?

It is an enzyme substance, similar to a vitamin, that aids organ and muscle cells and is purported to improve energy and slow the aging process. Research is being done to determine if it aids heart, lung, or Huntington disease patients.

28. What is the meaning of the term "umami"?

Science tells us that we can determine four tastes: salty, sweet, sour, and bitter. Umami is the fifth taste. It senses the presence of glutamate. This makes food taste richer, more savory, well-rounded, and more full-bodied.

29. Is any DDT used in the United States?

The DDT found in the U.S. comes in on fruits and vegetables from Mexico. Sometimes it drifts across the border from Mexican crop spraying or is in the water supply in the border states. Mexico continued to use this chemical to kill mosquitoes that cause malaria. They have now agreed to phase out its use over the next ten years.

30. What is pycnogenol that is so popular in health food stores?

It is the antioxidant isolated from the bark of a pine tree by Masquelier, a French chemist. This was about 50 years ago. The active principal is OPC, which is found in grape seeds.

31. What is the good ingredient in green tea?

Green tea leaves contain an easily absorbable natural antioxidant called polyphenols or catechins. Recent research indicates that it may be a protective factor against some fatal diseases, including cancer.

32. What are glucosamine sulfate and chondroitin sulfate?

These are synthetic versions of compounds the body makes to stimulate the growth of cartilage. The theory is that these compounds rebuild cartilage and reduce the symptoms of arthritis. It is an acceptable treatment in veterinary medicine, but the Arthritis Foundation and the American College of Rheumatology have not recommended it for humans. The Food and Drug Administration has not approved it to treat arthritis. At this time, they are sold as nutritional supplements, and therefore do not need evidence that they are safe and effective.

CHEMICALS

ABACA

Products and Uses: An extremely strong fiber obtained from a tree in the banana family. The fibers are woven into ropes and twines. Since it retains its strength when wet, it has many nautical applications.

Precautions: Harmless when consumed. Product is combustible. See marijuana entry for a related product.

Synonyms: MANILA PAPER ✦ HEMP ✦

ABSINTHE

Products and Uses: A green bitter oil used in liqueurs or vermouth as a flavoring. It is derived from wormwood. In early times it contained many heavy metals although today's product does not.

Precautions: Moderately toxic by swallowing. An allergen that may cause contact dermatitis (skin rash and irritation).

Synonyms: CAS 8022-37-5 ✦ ARTEMISIA OIL ✦ WORMWOOD

ACACIA

Products and Uses: An ingredient in tablet medications. It is a binder and a thickener in confectioneries, foods, cosmetics, adhesives, and textile printing.

Precautions: Harmless in finished product, but workers should avoid inhalation and contact, which could cause allergies, skin irritations, and possibly birth defects.

Synonyms: CAS 9000-01-5 ✦ ACACIA DEALBATA GUM ✦ ACACIA SENEGAL ✦ ACACIA SYRUP ✦ ARABIC GUM ✦ AUSTRALIAN GUM ✦ GUM

OVALINE ✦ GUM SENEGAL ✦ INDIAN GUM ✦ STARSOL NO. 1 ✦ WATTLE GUM

ACAROID RESIN

Products and Uses: Yacca tree gum material is used in inks and varnishes.

Precautions: Label directions must be observed.

Synonyms: GUM ACCROIDE ✦ YACCA GUM

ACENAPTHENE

Products and Uses: An additive in garden chemicals, dyes, and plastics and used as insecticide or fungicide.

Precautions: Derived from coal tar. A known carcinogen (causes cancer) and allergen.

Synonyms: ETHYLENENAPTHALENE ✦ 1,8-DIHYDROACENAPTHALENE

ACESULFAME POTASSIUM

Products and Uses: The newest artificial sweetener in beverage mixes, chewing gum, instant coffee, tea, dry dairy products, dry puddings, desserts, and tabletop sweetener. Used as a nonnutritive sweetener (approximately 24% of the sweetener market).

Precautions: FDA approves use.

Synonyms: CAS 55589-62-3 ✦ ACESULFAME K ✦ SUNETTE ✦ SWEET ONE

ACETIC ACID

Products and Uses: Commonly used in baked goods, catsup, cheese, chewing gum, condiments, dairy products, fats, gravies, mayonnaise, meat products, oil, pickles, poultry, relishes, salad dressings, and sauces, and in skin-bleaching cosmetics and hair-coloring products. Used as a flavor enhancer, pickling agent, and solvent.

Precautions: In its pure form it is moderately toxic when swallowed or breathed. Also considered a strong irritant to skin, tissue, and eyes. In excessive quantities when consumed it could cause reproductive effects (could cause infertility or sterility or birth defects). FDA states it is GRAS (generally recognized as safe) for the general population at levels normally consumed.

Synonyms: CAS 64-19-7 ✦ ETHANOIC ACID ✦ VINEGAR ACID ✦
METHANECARBOXYLIC ACID ✦ ETHYLIC ACID

ACETONE

Products and Uses: Utilized in nail polish, nail polish remover, markers, airplane glue solvent, fabric cement, cleaning fluids, paint, varnish, and lacquer as a solvent and to clean and dry.

Precautions: Flammable, dangerous fire risk. Moderately toxic by swallowing and breathing. A skin and severe eye irritant. Narcotic effect when inhaled in concentrated form. Can cause coma, kidney damage, and heart effects.

Synonyms: CAS 67-64-1 ✦ DIMETHYLKETONE ✦ 2-PROPANONE ✦ KETONE PROPANE ✦ METHYL KETONE ✦ PROPANONE ✦ PYROACETIC ACID ✦ PYROACETIC ETHER

ACETYL VALERYL

Products and Uses: Frequently used in beverages, ice cream desserts, candy, bakery goods, gum, cheese, butter, fruit, berries, nuts, and rum, as a flavoring.

Precautions: Harmless when used for intended purpose.

Synonyms: HEPTADIONE-2,3

ACID HYDROLYZED PROTEINS

Products and Uses: Commonly included in bologna, salami, sauces, and stuffing as a flavoring agent.

Precautions: USDA approves use by the general population in normal amounts. Must not be consumed by those with peanut/protein allergy.

Synonyms: HYDROLYZED MILK PROTEIN ✦ HYDROLYZED PLANT PROTEIN ✦ HYDROLYZED VEGETABLE PROTEIN

ACRYLAMIDE

Products and Uses: Applications include adhesives, textiles, and permanent-press fabrics as a coating and conditioner, also in soil conditioners.

Precautions: Suspected human carcinogen (may cause cancer). Toxic by skin absorption and an irritant to skin, nose, and throat.

Synonyms: CAS 79-06-1 ✦ ACRYLIC AMIDE ✦ ETHYLENECARBOXAMIDE ✦ PROPENAMIDE

ACTIVATED CARBON

Products and Uses: A decolorizing, odor-removing, purification agent, and taste remover in fats, juices, sherries, and wine.

Precautions: GRAS (generally recognized as safe).

Synonyms: CAS 64365-11-3 ✦ CHARCOAL, ACTIVATED

ADIPIC ACID

Products and Uses: Utilized in baked goods, baking powder, beverages (nonalcoholic), condiments, dairy product analogs, desserts (frozen dairy), drinks (powdered), fats, gelatins, gravies, margarine, meat products, oil, oil (edible), puddings, relishes, snack foods, and vegetables (canned) as a flavoring agent, leavening agent, neutralizing agent, and additive. Also used in the manufacture of nylon and polyurethane foam and lubricants.

Precautions: In pure form in large amounts it is moderately toxic and a severe eye irritant. However, harmless when used for intended purposes in limited amounts.

Synonyms: CAS 124-04-9 ✦ ACIFLOCTIN ✦ ACINETTEN ✦ ADILACTETTEN ✦ ADIPINIC ACID ✦ 1,4-BUTANEDICARBOXYLIC ACID ✦ 1,6-HEXANEDIOIC ACID ✦ KYSELINA ADIPOVA (CZECH) ✦ MOLTEN ADIPIC ACID

AFLATOXIN

Products and Uses: Naturally occurring toxin found in peanuts, walnuts, pistachios, pecans, almonds, and Brazil nuts. Also could be present in the mold or sprouting growth on grains. A natural contaminant. Prevention of mold growth is the best protection.

Precautions: A confirmed human carcinogen (causes cancer) and animal tumorigen (causes tumor growth). A human poison by swallowing. It cannot be totally eliminated. FDA sets allowable limitations.

Synonyms: CAS 1402-68-2

AGAR

Products and Uses: A preservative and gelling agent derived from red seaweed used in beverages, baked goods and mixes, candy, confections, frostings, glazes, jellied meats, dental impression material, cosmetics, and laxatives. The agar product used in labs for growth of bacteria, fungi, and so on, is mixed with blood or beef extract.

Precautions: A possible allergen. FDA states GRAS (generally recognized as safe).

Synonyms: CAS 9002-18-0 ✦ AGAR-AGAR ✦ AGAR AGAR FLAKE ✦ AGAR-AGAR GUM ✦ BENGAL GELATIN ✦ BENGAL ISINGLASS ✦ CEYLON ISINGLASS ✦ CHINESE ISINGLASS ✦ DIGENEA SIMPLEX MUCILAGE ✦ GELOSE ✦ JAPAN AGAR ✦ JAPAN ISINGLASS ✦ LAYOR CARANG

ALDICARB

Products and Uses: A pesticide used on many products such as beans, cotton, soybeans, potatoes, peanuts, citrus, sweet potatoes, coffee, and bananas. Has frequently been found in drinking water as a contaminant.

Precautions: At low doses it lowers the immune system of those who drink contaminated water. At high doses it disrupts the nervous system.

Synonyms: CAS 116-06-3 ✦ ALDECARB ✦ AMBUSH ✦ 2-METHYL-2-(METHYLTHIO)PROPANALO((METHYLAMINO)CARBONYL)OXIME ✦ TEMIK

ALDRIN

Products and Uses: Applications include insecticides and agricultural products such as fertilizers, herbicides, and fungicides.

Precautions: A poison by swallowing and skin contact. It has harmful effects on the human body by swallowing. When swallowed it causes confusion, tremors, nausea, or vomiting. Continued exposure causes liver damage. A human mutagen (changes inherited characteristics). Possibly causes cancer in humans.

Synonyms: CAS 309-00-2 ✦ ALDREX ✦ ALDREX 30 ✦ ALDRIN ✦ ALDRINE (FRENCH) ✦ ALDRITE ✦ ALDROSOL ✦ ALTOX ✦ DRINOX ✦ HHDN ✦ OCTALENE ✦ SEEDRIN

ALFALFA

Products and Uses: Commonly known as the dried leaf used as fodder for animals. For humans the herb is available in teas, tablets, drinks, and so on. Used as a folk treatment for arthritis, diabetes, and preventing cholesterol absorption.

Precautions: Moderate use is rarely harmful. A possible allergen. Individuals with Systemic Lupus Erythematous (SLE) must avoid.

Synonyms: MEDICAGO SATIVA ✦

ALGAE

Products and Uses: The plant growth from water in ponds and lakes. There are a number of forms. Many benefits have been attributed to these including weight loss, energy gains, lower cholesterol, and so on. They provide iron, calcium, and potassium as well.

Precautions: Some algae produce liver toxins.

Synonyms: SPIRULINA BLUE-GREEN ALGAE ✦ APHANIZOMENON FLOS-AQUAE ✦ CHLORELLA

ALGINIC ACID

Products and Uses: Various applications are in soups, antacids, ice cream, toothpaste, and cosmetics, as a thickener, emulsifier, or stabilizer. Also a waterproofing agent for concrete.

Precautions: FDA states it is GRAS (generally recognized as safe).

Synonyms: CAS 9005-32-7 ✦ KELACID ✦ LANDALGINE ✦ NORGINE ✦ PLOYMANNURONIC ACID ✦ SAZZIO

ALKALOID

Products and Uses: Examples are morphine, nicotine, quinine, codeine, caffeine, cocaine, and strychnine. Caffeine is added to beverages and is found naturally in coffee and tea. Nicotine is naturally occurring in tobacco products and is added to pesticides for its toxic qualities. See nicotine, codeine, and so forth. for related information.

Precautions: Depending on dosage, these can be poisonous. Can also be an allergenic.

Synonyms: ALKALOID SALTS

ALKYL POLYGLYCOSIDES

Products and Uses: Utilized in detergents, soaps, and shampoos; a surfactant (surface active agent) that separates the soil from the item being cleaned.

Precautions: Reported to be biodegradable and less irritating to skin and eyes than similar chemicals.

Synonyms: GLUCOPON

ALLSPICE

Products and Uses: Commonly utilized in cakes, fruit pies, mincemeat, plum puddings, sauces, many food items, and soups. A flavoring agent derived from berries of the tree.

Precautions: A sensitizer that after repeated contact can cause dermatitis. Considered moderately toxic in large amounts.

Synonyms: PIMENTA OIL ✦ PIMENTA BERRIES OIL ✦ PIMENTO OIL

ALLYL ALCOHOL

Products and Uses: An herbicide (kills plants and vegetation) in garden and agricultural products. Also used in military poison gas and the manufacture of plastics, perfumes, and medications.

Precautions: A poison by breathing, swallowing, and skin contact; also a skin and severe eye irritant.

Synonyms: CAS 107-18-6 ✦ ALLYLIC ALCOHOL ✦ 3-HYDROXYPROPENE ✦ ORVINYLCARBINOL ✦ PROPENOL ✦ PROPENYL ALCOHOL ✦ WEED DRENCH

ALLYL ISOTHIOCYANATE

Products and Uses: Applications include baked goods, condiments, horseradish flavor (imitation), meat, mustard oil (artificial), pickles, salad dressings, and sauces as a flavoring agent. Also found in mustard plasters and over-the-counter medications and in fumigants.

Precautions: FDA approves use as it is considered safe for the general population at levels normally consumed; possibly toxic in high doses. Suspected carcinogen (might cause cancer). An eye irritant and allergen. May cause contact dermatitis.

Synonyms: CAS 57-06-7 ✦ AITC ✦ ALLYL ISORHODANIDE ✦ ALLYL ISOSULFOCYANATE ✦ ALLYL MUSTARD OIL ✦ ALLYL SEVENOLUM ✦ ALLYL THIOCARBONIMIDE ✦ ARTIFICIAL MUSTARD OIL ✦ CARBOSPOL ✦ 3-ISOTHIOCYANATO-1-PROPENE ✦ MUSTARD OIL ✦ OLEUM SINAPIS VOLATILE ✦ 2-PROPENYL ISOTHIOCYANATE ✦ REDSKIN ✦ VOLATILE OIL OF MUSTARD

ALLYL SULFIDE

Products and Uses: Occurs naturally in garlic and horseradish although used as a fruit flavoring in beverages, ice cream, condiments, and meats.

Precautions: An irritant to skin, eyes, nose, and throat.

Synonyms: CAS 592-88-1 ✦ ALLYL MONOSULFIDE ✦ DIALLYL SULFIDE ✦ DIALLYL THIOETHER ✦ OIL GARLIC ✦ THIOALLYL ETHER

ALLYL TRICHLOROSILANE

Products and Uses: Used in boat and car repair products for fiber glass finishes.

Precautions: A fire hazard. Possible skin and eye irritant.

Synonyms: TRICHLOROALLYLSILANE

ALOE VERA

Products and Uses: Aloe is a succulent plant in the lily family. The leaf contains a gel-like substance used for skin lotions, creams, ointments, soaps, makeup, moisturizers, and over-the-counter (OTC) burn medications. It is a popular cosmetic and pharmaceutical ingredient. Frequently sold as a drink and mixed with water and juices. Could produce a laxative effect when consumed.

Precautions: Could cause allergic reaction in susceptible individuals. Believed to be beneficial and healing on external skin burns.

Synonyms: LALOI ✦ SABILA ✦ SEMPERVIVUM ✦ SINKLE BIBLE ✦ STAR CACTUS ✦ ZABILA ✦ ZAVILA

ALOIN

Products and Uses: Utilized in medicines such as OTC (over-the-counter) laxatives; used as a purgative.

Precautions: A natural product with no known toxicity. Derived from aloe vera plant.

Synonyms: BARBALOIN

ALPHA HYDROXY ACIDS

Products and Uses: Derived from fruit, milk, and natural sources. Primarily lactic and glycolic acids that work as a special moisturizer that restores smoothness to skins roughened, chapped, or thickened from climate or disease.

Precautions: Slightly stinging upon first application. A fruit-based acid safe for pregnant women and all skin types.

Synonyms: AHA

ALPHA TOCOPHEROL

Products and Uses: Used in baby lotions, hair cosmetic products, and deodorant products as an antioxidant (slows reaction with oxygen) to prevent oils from becoming rancid; as a nutrient.

Precautions: Harmless when used for intended purposes.

Synonyms: VITAMIN E

ALUMINA TRIHYDRATE

Products and Uses: An ore derived from bauxite used in glazes, ceramic coatings, paper coatings, cosmetics, glass, ceramic hobby production, paper correction liquids, mattresses, makeup products filler, and flame retardant.

Precautions: Swallowing causes fever and gastrointestinal (stomach) effects. Can cause mutagenic effects (can change inherited characteristics).

Synonyms: CAS 21645-51-2 ✦ ALUMIGEL ✦ ALUMINA HYDRATE ✦ ALUMINIC ACID ✦ ALUMINUM HYDRATE ✦ ALUMINUM HYDROXIDE GEL ✦ ALUMINUM OXIDE HYDRATE ✦ AMPHOJEL ✦ HIGILITE ✦ LIQUIGEL

ALUMINUM

Products and Uses: Found as a powder in paints, protective coatings, used as foil packaging, ointment tubes, cooking ware, and for corrosion resistance.

Precautions: Fine powder forms flammable and explosive mixtures in air. Some suspicion, but no scientific evidence, that aluminum cookware, foil, or even antacid tablets containing aluminum, cause Alzheimer's disease.

Synonyms: CAS 7429-90-5 ✦ ALUMINUM DEHYDRATED ✦ ALUMINUM FLAKE ✦ ALUMINUM POWDER ✦ METANA ALUMINUM PASTE

ALUMINUM CHLORIDE

Products and Uses: Utilized in antiperspirants and deodorants to prevent perspiration.

Precautions: Possible allergen and irritant, also moderately toxic by swallowing.

Synonyms: CAS 7446-70-0 ✦ ALUMINUM CHLORIDE ANHYDROUS ✦ ALUMINUM CHLORIDE SOLUTION ✦ ALUMINUM TRICHLORIDE ✦ PEARSALL ✦ TRICHLOROALUMINUM

ALUMINUM DEXTRAN

Products and Uses: Found in packaging materials, makeup products, and various food additives as an anticaking agent, binder in cosmetic compressed powders, emulsifier, and stabilizer.

Precautions: Must conform to FDA specifications for salts or fats or fatty acids derived from edible oils.

Synonyms: CAS 7047-84-9 ✦ ALUMINUM MONOSTEARATE ✦ ALUMINUM STEARATE ✦ STEARIC ACID, ALUMINIUM SALT

ALUMINUM OXIDE

Products and Uses: A food additive and antacid; also a dispersing agent for white coloring matter.

Precautions: In gross amounts it is a possible carcinogen (cancer causing substance). Breathing of fine particles can cause lung damage. Note: The gemstones ruby and sapphire are aluminum oxide colored by traces of chromium and cobalt in another form.

Synonyms: CAS 1344-28-1 ✦ ALUMINA ✦ ALUMITE ✦ ALUNDUM ✦ ACTIVATED ALUMINUM OXIDE ✦ ALUMINUM SESQUIOXIDE

ALUMINUM PHOSPHATE

Products and Uses: Typically used in dental cements, cosmetics, paints and varnishes as a gelling agent, an antacid, and a cement component.

Precautions: These solutions are corrosive to tissue.

Synonyms: CAS 7784-30-7 ✦ ALUMINOPHOSPHORIC ACID ✦ ALUMINUM ACID PHOSPHATE ✦ ALUPHOS ✦ MONOALUMINUM PHOSPHATE ✦ PHOSPHALUGEL

ALUMINUM PHOSPHIDE

Products and Uses: A fumigant or insecticide on brewer's corn, grits, brewer's malt, and brewer's rice.

Precautions: A human poison by breathing and swallowing large amounts. Dangerous fire risk. On EPA extremely hazardous substance list.

Synonyms: CAS 20859-73-8 ✦ AIP ✦ AL-PHOS ✦ ALUMINUM FOSFIDE (DUTCH) ✦ ALUMINUM MONOPHOSPHIDE ✦ CELPHIDE ✦ CELPHOS ✦ DELICIA ✦ FUMITOXIN

ALUMINUM SULFATE (2:3)

Products and Uses: Prevents skin irritation in antiperspirants that contain aluminum chloride. Also in after-shave lotions, and in styptic pencils. Put in animal glue, packaging materials, pickle relish, pickles, potatoes, and shrimp packs as a firming agent.

Precautions: FDA states GRAS (generally recognized as safe) when used for intended purpose. Moderately toxic by swallowing.

Synonyms: CAS 10043-01-3 ✦ ALUMINUM TRISULFATE ✦ CAKE ALUM ✦ DIALUMINUM SULFATE ✦ SULFURIC ACID, ALUMINUM SALT ✦ ALUM ✦ DIALUMINUM TRISULFATE

AMALGAM

Products and Uses: A mixture of mercury with silver tin alloy in dental fillings, and for silvering mirrors. It is also a binder for precious metals such as gold and silver.

Precautions: Mercury is highly toxic by skin absorption or breathing of fumes.

Synonyms: MERCURY SILVER TIN ALLOY

AMMONIA GAS

Products and Uses: Commonly used in fertilizers, yeast nutrients, film developers, and refrigerants.

Precautions: Breathing of concentrated fumes may be fatal. Explosive compounds form in contact with silver or mercury.

Synonyms: CAS 7664-41-7 ✦ AMMONIAC ✦ AMMONIA ANHYDROUS ✦ ANHYDROUS AMMONIA ✦ SPIRIT OF HARTSHORN

AMMONIA, AROMATIC SPIRITS

Products and Uses: A 10% mixture of ammonia in alcohol. Used in inhalers and as a respiratory stimulant.

Precautions: Irritant to nose and throat.

Synonyms: 10% AMMONIA IN ALCOHOL

AMMONIA WATER

Products and Uses: A neutralizer in hair permanent wave solution, hair straighteners, hair coloring agents, and skin creams; also a hair texture changer.

Precautions: Toxic when breathed. Severe irritant to eyes, nose, and throat.

Synonyms: AMMONIA, LIQUID

AMMONIUM BICARBONATE

Products and Uses: Found in baked goods as a leavening agent. Varied uses include permanent wave solutions and medications for intestinal gas.

Precautions: FDA states GRAS (generally recognized as safe) when used for intended purpose.

Synonyms: CAS 1066-33-7 ✦ AMMONIUM ACID CARBONATE ✦ AMMONIUM HYDROGEN CARBONATE ✦ CARBONIC ACID ✦ MONOAMMONIUM CARBONATE ✦ MONOAMMONIUM SALT

AMMONIUM BORATE

Products and Uses: A component in outdoor and garden herbicide products; also used for fireproofing of fabrics.

Precautions: Package directions must be followed carefully.

Synonyms: AMMONIUM BIBORATE

AMMONIUM CARBONATE

Products and Uses: Major ingredient in smelling salts, also in baked goods, baking powder, caramel, gelatins, puddings, and wine. Utilized as a buffer, leavening agent, neutralizing agent, and yeast nutrient.

Precautions: FDA states GRAS (generally recognized as safe). Safe for the general population at levels normally consumed.

Synonyms: CAS 506-87-6 ✦ CARBONIC ACID, AMMONIUM SALT ✦ DIAMMONIUM CARBONATE ✦ CRYSTAL AMMONIA ✦ AMMONIUM SESQUICARBONATE ✦ HARTSHORN

AMMONIUM CHLORIDE

Products and Uses: The purpose in baked goods is as a dough conditioner; in condiments and relishes as a flavor enchancer; also in washing powders. It is sprayed on snow in ski areas to prevent melting.

Precautions: FDA states GRAS (generally recognized as safe) when used for intended purposes in baked goods. Moderately toxic if consumed in gross amounts. A severe eye irritant.

Synonyms: CAS 12125-02-9 ✦ AMMONIUMCHLORID (GERMAN) ✦ AMMONIUM MURIATE ✦ CHLORID AMONNY (CZECH) ✦ SAL AMMONIA ✦ SAL AMMONIAC

AMMONIUM HYDROXIDE

Products and Uses: Another versatile chemical found in baked goods as a leavening agent, in caramel, cheese, fruits (processed), and puddings. Also in ammonia soaps, for fireproofing woods, in detergents, household cleaners, and hair dyes.

Precautions: FDA states GRAS (generally recognized as safe) when used for intended purposes. A poison and severe irritant to mouth and throat when large amounts are swallowed.

Synonyms: CAS 1336-21-6 ✦ AMMONIA AQUEOUS ✦ AMMONIA SOLUTION ✦ AQUA AMMONIA

AMMONIUM NITRATE

Products and Uses: Commonly found in agricultural products such as fertilizers, herbicides, and insecticides. Used in manufacture of fireworks and matches. It is in the freezing mixture ingredients in picnic coolers.

Precautions: May explode under confinement and when exposed to high temperatures, but not readily detonated.

Synonyms: CAS 6484-52-2 ✦ NITRIC ACID, AMMONIUM SALT ✦ NORWAY SALTPETER

AMMONIUM PHOSPHATE

Products and Uses: Utilized for flameproofing of wood, paper, and textiles. Used for vegetation coating to retard forest fires; also it is added to matches and candles to prevent afterglow and smoking. It is in ammoniated dentifrices and also in fertilizers.

Precautions: Harmless when used for intended purposes.

Synonyms: DIAMMONIUM HYDROGEN PHOSPHATE ✦ DIAMMONIUM PHOSPHATE ✦ DAP

AMYL ACETATE

Products and Uses: A solvent in nail lacquers, polish remover, and leather polish. A perfume odorant, and artificial fruit flavoring, as well as use in manufacturing of plastics and inks and antibiotics.

Precautions: Human systemic effects by breathing. Mildly toxic. Skin and eye irritation. Flammable, dangerous fire hazard.

Synonyms: CAS 628-63-7 ✦ ACETIC ACID, AMYL ESTER ✦ AMYL ACETIC ESTER ✦ PEAR OIL ✦ BANANA OIL ✦ BIRNENOEL ✦ PENT-ACETATE ✦ PENTYL ACETATE

AMYL FORMATE

Products and Uses: Used in films, coatings, leather, and perfume as solvent, odorant, and flavoring.

Precautions: Flammable, dangerous fire risk; also toxic by swallowing and breathing. Skin irritant.

Synonyms: CAS 638-49-3 ✦ PENTYL FORMATE

ANETHOLE

Products and Uses: Utilized in perfume, anise flavor for dentifrice, and licorice candy, as a flavoring agent.

Precautions: Possible irritant to mouth, nose, and throat; a possible skin allergen.

Synonyms: ANISE OIL ✦ ANISE CAMPHOR ✦ p-PROPENYLANISOLE ✦ p-METHOXYPROPENYLBENZENE

ANTHRALIN

Products and Uses: Found in medications for the treatment of psoriasis on extremities. Used as an ointment; not for severe cases.

Precautions: Very irritating. Not for use on scalp or near eyes.

Synonyms: 1,8,9-ANTHRACENETRIOL ✦ 1,8-DIHYDROXYANTHRANOL

ANTIBIOTIC

Products and Uses: A chemical substance produced by microorganisms that has the ability to inhibit the growth of other microorganisms or destroy them. Synonyms are tyrothricin, bacitracin, polymxin, actinomycin, streptomycin, chloramphenicol, tetracycline. Certain antibiotics are used as food additives to inhibit the growth of bacteria and fungi. These are nisin, pimaricin, nystatin, tylosin.

Precautions: A possible allergen. Overuse of antibiotics can lead to the development of resistant strains of antibiotics. Some antibiotics enter humans via the food chain. Antibiotics from cattle that have been treated with antibiotic medication and slaughtered too early can reach the consumer. The same can happen with milk cows that have had udder infections.

Synonyms: AS ABOVE.

ANTIMONY TRISULFIDE

Products and Uses: In matches and camouflage paints. As a coloring agent in fireworks.

Precautions: Blood and gastrointestinal (stomach and digestive) effects by breathing. Possible carcinogen (causes cancer).

Synonyms: CAS 1345-04-6 ✦ ANTIMONOUS SULFIDE ✦ ANTIMONY
GLANCE ✦ ANTIMONY ORANGE ✦ ANTIMONY SULFIDE ✦
LYMPHOSCAN ✦ NEEDLE ANTIMONY

ANTIOXIDANTS

Products and Uses: Compounds such as vitamin C, E and beta carotene. Antioxidants control "free radicals," which damage cells through oxidation. These vitamins are abundant in fruits and vegetables; therefore the beneficial effects of eating fruits and vegetables is attributed to the antioxidants.

Precautions Some marketers suggest all antioxidants can prevent cancer and heart disease, but the evidence from controlled clinical trials is mixed and suggests that different antioxidants have different effects.

Synonyms: VITAMIN C ✦ VITAMIN E ✦ VITAMIN A

ANTISEPTIC

Products and Uses: A substance that retards or stops the growth of microorganisms. Examples are alcohol, boric acid, barates, acriflavine, menthol, hydrogen peroxide, hypochlorites, iodine, mercuric chloride, phenol, hexachlorophene, and quaternary ammonium compounds.

Precautions: Could possibly be corrosive or toxic, depending on application.

Synonyms: ANTIBACTERIAL ✦ ANTIMICROBIAL

ARSENIC COMPOUNDS

Products and Uses: Highly toxic compounds used in pesticides and fungicides to treat lumber, in metal alloys, some glass production, medications for blood and skin disorders, and even in cosmetic products. Lethal dose for adults is only one-half gram.

Precautions: Irritating to skin and eyes, causes lung and liver cancers, poison if consumed.

Synonyms: NONE LISTED.

ARTEMISIA OIL

Products and Uses: In beverages (alcoholic) as a bitter flavoring agent.

Precautions: FDA approves use at moderate levels to accomplish the intended effect. Moderately toxic by swallowing. An allergen. Habitual users develop "absinthism" with mental and physical deterioration from overuse. May cause a contact dermatitis.

Synonyms: CAS 8022-37-5 ✦ ABSINTHIUM ✦ ARTEMISIA OIL (WORMWOOD) ✦ OIL, ARTEMISIA

ARTIFICIAL COLORINGS

Products and Uses: Usually found in low nutritional, high sugar content foods, candy, carbonated drinks, ice sticks, and gelatins. Colorings are added when natural fruit colorings are absent. Not required to be listed on labels.

Precautions: Avoid when possible. Problems can range from allergic reactions to being carcinogenic (cancer causing). See individual coloring names.

Synonyms: SEE SPECIFIC COLORS.

ARTIFICIAL FLAVORINGS

Products and Uses: Frequently found in low nutritional, high sugar content foods, candy, carbonated drinks, ice sticks, gelatins, and breakfast cereals. Flavorings are added when real fruit, or natural flavor, are absent. Recently, some research states that there may be benefits in consuming artificial flavorings. Flavorings contain salicylates, a chemical cousin of aspirin, and aspirin lowers the risk of heart attack by preventing blood clots. Most people are said to consume the equivalent of one baby aspirin a day from artificial flavorings.

Precautions: Flavorings that occur in nature are usually safe. Artificial flavorings are usually less desirable.

Synonyms: SEE SPECIFIC FLAVORS.

ARTIFICIAL SNOW

Products and Uses: Spray for winter and Christmas displays and decorations.

Precautions: Flammable. Mildly toxic by swallowing.

Synonyms: COPOLYMER OF BUTYL METHACRYLATE ✦ COPOLYMER OF ISOBUTYL METHACRYLATE

ASBESTOS

Products and Uses: Mineral fiber used in fireproof materials, brake linings, gaskets, roofing compositions, paint filler, and filters.

Precautions: A carcinogen (causes cancer). Usually four to seven years of exposure are required before serious lung damage results.

Synonyms: CAS: 1332-21-4 ✦ AMIANTHUS ✦ AMOSITE ✦ AMPHIBOLE ✦ FIBROUS GRUNERITE ✦ SERPENTINE

ASCORBIC ACID

Products and Uses: In beverages, potato flakes, breakfast foods, beef (cured), meat food products, pork (cured and fresh), sausage, and wine. A dietary supplement, nutrient, preservative, and antioxidant (slows reaction with oxygen) to increase shelf life.

Precautions: FDA states GRAS (generally recognized as safe). Safe for the general public in amounts normally consumed.

Synonyms: CAS: 50-81-7 ✦ l-ASCORBIC ACID ✦ VITAMIN C ✦ ASCORBUTINA ✦ CEVITAMIC ACID ✦ CEVITAMIN ✦ VITACIN ✦ VITAMISIN ✦ VITASCORBOL

ASPARTAME

Products and Uses: Artificial sweetener in beverages (carbonated and dry base), breath mints, cereals, chewable multivitamins, chewing gum, coffee (instant dry base), frozen stick confections, dairy product topping, fruit flavored drinks and ades, fruit juice based drinks, puddings, and tea. As a flavor enhancer, sugar substitute (approximately 71% of market).

Precautions: FDA approves use at moderate levels. Made from two amino acids. It is equal to sucrose in calories, but it is 200 times sweeter. Therefore, the amount needed to sweeten food is negligible. Could cause allergic dermatitis. Caused reproductive effects (infertility or sterility or birth defects) in animal studies. Aspartame contains a chemical called phenylanine that is dangerous to people with the disease phenylketonuria (PKU).

Synonyms: CAS: 22839-47-0 ✦ EQUAL ✦ CANDEREL ✦ DIPEPTIDE SWEETNER ✦ METHYL ASPARTYLPHENYLALANATE ✦ NUTRASWEET ✦ ASPARTYLPHENYLALANINE METHYL ESTER ✦ SWEET DIPEPTIDE

ASPHALT

Products and Uses: Road, roof, and foundation coatings. In rubber coverings for electrical wiring and pipes.

Precautions: A suspected carcinogen (causes cancer). Moderately irritating to skin. Can cause dermatitis. Fumes can be irritating to nose, eyes, and lungs.

Synonyms: CAS: 8052-42-4 ✦ ASPHALTUM ✦ BITUMEN ✦ MINERAL PITCH ✦ PETROLEUM PITCH ✦ ROAD ASPHALT ✦ ROAD TAR

ASPIRIN

Products and Uses: An OTC (over-the-counter) analgesic, anti-inflammatory, antipyretic (to reduce fever). Certain foods contain salicylates, the active ingredient in aspirin. They may afford the same benefits. They are raisins, prunes, dates, dried currants, raspberries, blueberries, cherries and Granny Smith apples.

Precautions: An allergen. May cause bleeding. A 10 g dose may be fatal. Sometimes prescribed to prevent heart attacks because of anticoagulant action. Recent studies indicate it may be a colon cancer preventative. Should be avoided before surgical procedure to avoid excessive bleeding. Should not be taken by children because of Reyes syndrome implication.

Synonyms: ACETYLSALICYLIC ACID ✦ o-ACETOXYBENZOIC ACID

ASTRAGALUS

Products and Uses: A dried root that is supplied as a tea, in capsules, tablets, tinctures, and so forth. In Chinese medicine it is a folk remedy for a variety of ailments including head colds, flu, stomach ulcers, and diabetes. Used to treat cancer patients in China who are undergoing chemotherapy and radiation treatment. It also is reported to have desirable cardiovascular effects.

Precautions: No undesirable side effects are reported.

Synonyms: ASTRAGALUS MEMBRANACEUS

BAKELITE

Products and Uses: A plasticlike material developed in 1907 by Belgium-born L. H. Bakeland. It is a phenol-formaldehyde, a hard heat-resistant material used for toaster handles, radio knobs, and so on.

Precautions: Harmless in normal use.

Synonyms: NONE KNOWN.

BAKER'S YEAST EXTRACT

Products and Uses: Utilized in cheese spread flavorings, frozen desserts, salad dressings, cheese flavored snack dips, soups, sour cream, and wines. As an emulsifier, flavoring agent, nutrient supplement, stabilizer, and thickener yeast food.

Precautions: FDA states GRAS (generally recognized as safe) when used for intended purpose. Limited to 5% in salad dressing.

Synonyms: AUTOLYZED YEAST EXTRACT ✦ BAKER'S YEAST GLYCAN

BAKING SODA

Products and Uses: Applications in food, mouthwashes, antacids, skin powders, and bath salts. Used as a leavening agent or to adjust acidity in foods and in laundry or refrigerator as deodorizer.

Precautions: In rare cases there have been reports of people taking baking soda internally for indigestion after large meals and experiencing stomach rupture. This occurs because baking soda releases carbon dioxide and the stomach is too full, or there is damage as a result of ulcers or obstruction. It has high sodium content, which can contribute to high blood pressure, heart disease, and kidney problems. GRAS (generally recognized as safe).

Synonyms: SODIUM BICARBONATE

BALSAM OF PERU

Products and Uses: As a flavoring and fragrance for smoky, richly balsamic or cinnamic flavors. Derived from evergreen trees or shrubs. In chocolate manufacturing, expectorants, cough syrups, shampoo fragrances, and hair conditioners.

Precautions: An allergen. Combustible when heated.

Synonyms: BALSAM PERU OIL   ✦   PERUVIAN BALSAM

BARIUM THIOSULFATE

Products and Uses: Additive in luminous paints, matches, and varnishes as pigments, glazes, and protective coatings.

Precautions: Flammable. Toxic by swallowing and breathing.

Synonyms: BARIUM HYPOSULFITE

BATTERIES

Products and Uses: There are five main types: lead storage batteries for autos, motorcycles, lawnmowers, and so on; zinc chloride or zinc carbon batteries for calculators and clocks; alkaline batteries used in cameras and radios; nicklecadmium rechargable batteries that can be used repeatedly; and button cell batteries used in watches and hearing aids. Batteries are not energy efficient. Their manufacture requires 50 times more energy than they will produce.

Precautions: Lead storage, zinc, chloride, alkaline, rechargable and button cell types can contain mercury and cadmium, which are toxic metals. They should be treated as toxic waste.

Synonyms: NONE KNOWN.

BAUXITE

Products and Uses: The most common ore in earth's crust. Basis for aluminum and all aluminum products. Also used as a filler in rubber, paints, plastics, and cosmetics.

Precautions: When inhaled, powder can cause pulmonary fibrosis. Dust is flammable and explosive.

Synonyms: NONE KNOWN.

BAYOIL

Products and Uses: Fresh sweet, cineolic flavor used for bay rum, meats, soups, and stews. For fragrances and flavoring it is an ingredient that affects the taste or smell of final product.

Precautions: FDA states GRAS (generally recognized as safe) when used for intended purposes. Moderately toxic by swallowing large amounts.

Synonyms: BAYLEAF OIL ✦ LAUREL LEAF OIL ✦ MYRCIA OIL ✦ OIL OF BAY

BEESWAX

Products and Uses: Bees secrete this wax and use it to construct their cells. In candy (hard and soft), candy glaze, creams, ear plugs, lipstick, and church candles. Also used as a flavoring and in polishes, waxes, adhesives, and textile sizing.

Precautions: Combustible. A mild allergen to some people.

Synonyms: CAS: 8012-89-3 ✦ BEESWAX, WHITE ✦ BEESWAX, YELLOW

BENOMYL

Products and Uses: An after-harvest fungicide used on fruits such as apples, apricots, bananas, cherries, plums, pears, raisin, and also tomato products.

Precautions: Highly toxic to earthworms and some fish. Poison by swallowing. Mildly toxic by breathing. An animal teratogen (abnormal fetus development). Caused reproductive effects in animal studies (infertility or sterility or birth defects). A human mutagen (changes inherited characteristics). A skin irritant. EPA was ordered by courts to revoke approval as food additive in 1992.

Synonyms: CAS: 17804-35-2 ✦ BNM ✦ BENLATE ✦ FUNDASOL

BENTONITE

Products and Uses: A smectite clay from volcanic ash. Used in foods such as fruit flavorings, liquors, ice cream, baked goods, and chewing gum. Also in cosmetics, facial makeup, facial masks, and used to clarify and stabilize wines. It can be a colorant, pigment, stabilizer, or thickener. Also in water softeners and in drilling muds.

Precautions: FDA states GRAS (generally recognized as safe) when used for intended purposes.

Synonyms: CAS: 1302-78-9 ✦ ALBAGEL PREMIUM USP 4444 ✦ HI-JEL ✦ MAGBOND ✦ MONTMORILLONITE ✦ PANTHER CREEK BENTONITE ✦ SOUTHERN BENTONITE ✦ TIXOTON ✦ VOLCLAY ✦ WILKINITE

BENZALDEHYDE

Products and Uses: Possesses the flavor and odor of bitter almonds; it is used in perfume, soaps, cosmetic oils, and dyes.

Precautions: An allergen. Has very mild local anesthetic properties. A skin irritant. Causes poisoning and convulsions in large doses.

Synonyms: CAS: 100-52-7 ✦ ARTIFICIAL ALMOND OIL ✦ BENZENECARBALDEHYDE ✦ BENZENECARBONAL ✦ BENZOIC ALDEHYDE

BENZENE

Products and Uses: An aromatic compound that is not in current use in consumer products. It occurs naturally in crude oil; also found in gasoline and petroleum mixes. Still found in old cosmetics, perfumes, nail polish remover, airplane glues, lacquers, dry-cleaning products, paint, spot remover, varnish, stain, and sealant. Was used for solvents, coatings, and various other uses.

Precautions: A confirmed human carcinogen (causes cancer), induces myeloid leukemia, Hodgkin's disease, and lymphomas by breathing. A human poison by breathing. Moderately toxic by swallowing. A severe eye and skin irritant. Effects on the body by swallowing and breathing are blood changes and increased body temperature. Effects are cumulative. A human mutagen (changes inherited characteristics). A narcotic. Benzene can penetrate the skin and cause poisoning. Products containing 5% or more by weight must be labeled "Danger: Vapor Harmful, Poison" with the skull and crossbones symbol. If it contains 10% or more, it must also state "Harmful or fatal if swallowed. Call physician immediately." This is the highest volume chemical produced in the U.S.

Synonyms: CAS: 71-43-2 ✦ BENZOL ✦ BENZOLE ✦ BENZOLENE ✦ BICARBURET OF HYDROGEN ✦ CARBON OIL ✦ COAL NAPHTHA ✦ MOTOR BENZOL ✦ NITRATION BENZENE ✦ PHENE ✦ PYROBENZOL ✦ PYROBENZOLE ✦ PHENYL HYDRIDE

BENZENEACETALDEHYDE

Products and Uses: A flavoring in products such as beverages, chewing gum, confections, gelatins, ice cream, maraschino cherries, and puddings.

Precautions: Moderately toxic by swallowing large amounts. A skin irritant.

Synonyms: CAS: 122-78-1 ✦ HYACINTHIN ✦ PAA ✦ PHENYLACETALDEHYDE ✦ PHENYLACETIC ALDEHYDE ✦ PHENYLETHANAL ✦ α-TOLUALDEHYDE ✦ α-TOLUIC ALDEHYDE

BENZOCAINE

Products and Uses: In medicine (local anesthetic) and in sunburn medications. To numb or deaden skin to prevent pain.

Precautions: Toxic by swallowing.

Synonyms: CAS: 51-05-8 ✦ ANESTHESOL ✦ PROCAINE HYDROCHLORIDE ✦ ETHYL-p-AMINOBENZOATE HYDROCHLORIDE

BENZOIC ACID

Products and Uses: Naturally occuring in cinnamon, cloves (ripe), cranberries, plums, and prunes. Used in food preservatives, tobacco seasoning, flavors, perfumes, and toothpastes. Utilized orally as an antiseptic, diuretic, and expectorant and as an antimicrobial (antiseptic) agent, flavoring agent, and food preservative.

Precautions: Moderately toxic by swallowing gross amounts. Effects on the body by breathing are dyspnea (shortness of breath) and skin allergy. A severe eye irritant and skin irritant. GRAS (generally recognized as safe) with a limitation of 0.1% in foods.

Synonyms: CAS: 65-85-0 ✦ BENZENECARBOXYLIC ACID ✦ BENZENEFORMIC ACID ✦ BENZENEMETHANOIC ACID ✦ BENZOATE ✦ CARBOXYBENZENE ✦ DRACYLIC ACID ✦ PHENYL CARBOXYLIC ACID ✦ PHENYLFORMIC ACID ✦ RETARDEX ✦ SALVO

BENZOYL PEROXIDE

Products and Uses: Varied uses include as a bleaching agent for flour, fats, oils, and waxes; also in antiacne pharmaceuticals and cosmetics, cheese production, and in the embossing of vinyl flooring.

Precautions: A poison by swallowing large amounts. Can cause dermatitis. An allergen and eye irritant. A mutagen (changes inherited characteristics). Possible carcinogen (causes cancer) caused tumor growth in animals. GRAS (generally recognized as safe).

Synonyms: CAS: 94-36-0 ✦ ACETOXYL ✦ ACNEGEL ✦ BENOXYL ✦ BENZAC ✦ BENZOIC ACID, PEROXIDE ✦ BENZOPEROXIDE ✦ BENZOYL ✦ BENZOYL SUPEROXIDE ✦ CLEARASIL BENZOYL PEROXIDE LOTION ✦ CLEARASIL BP ACNE TREATMENT ✦ CUTICURA ACNE CREAM ✦ DEBROXIDE ✦ DRY AND CLEAR ✦ EPI-CLEAR ✦ FOSTEX ✦ OXY-5 ✦ OXY-10 ✦ OXYLITE ✦ OXY WASH

BENZYL ACETATE

Products and Uses: A colorless liquid with a floral aroma used in artificial jasmine, perfumes, soap perfume, lacquers, polishes, inks, and varnish remover.

Precautions: Highly toxic by breathing. Skin irritant.

Synonyms: CAS: 140-11-4 ✦ ACETIC ACID BENZYL ESTER ✦ ACETIC ACID PHENYLMETHYL ESTER ✦ BENZYL ETHANOATE

BENZYL ALCOHOL

Products and Uses: Utilized in perfumes, flavors, antibacterials, cosmetics, ointments, ballpoint pen ink, stencil ink, and hair dye preservative. Used as a flavoring, odorant, solvent, preservative, and pharmaceutical.

Precautions: Highly toxic. Poison by swallowing. Moderately toxic by breathing and skin contact. A moderate skin and severe eye irritant.

Synonyms: CAS: 100-51-6 ✦ BENZAL ALCOHOL ✦ BENZENECARBINOL ✦ BENZENEMETHANOL ✦ PHENOCARBINOL ✦ PHENYLMETHANOL ✦ PHENYL-METHYL ALCOHOL

BENZYL BENZOATE

Products and Uses: Commonly used in perfume musk, external medicines, plasticizer in nail polish, and miticide; as a fixative, solvent, and flavoring.

Precautions: Possible allergen. Irritant to eyes and skin.

Synonyms: CAS: 120-51-4 ✦ ASCABIN ✦ ASCABIOL ✦ BENYLATE ✦ BENZYL BENZENECARBOXYLATE ✦ BENZYL BENZFORMATE ✦ NOVOSCABIN ✦ VENZONATE

o-BENZYL-p-CHLOROPHENOL

Products and Uses: Utilized in soaps as a disinfectant.

Precautions: Highly toxic. An irritant. Combustible.

Synonyms: CHLOROPHENE ✦ SANTOPHEN ✦ SEPTIPHENE ✦ 4-CHLORO-α-PHENYL-o-CRESOL

BENZYL CINNAMATE

Products and Uses: Frequently found in fruit-scented perfumes and flavors as a flavoring and odorant. An ingredient that affects the taste or smell of the final product.

Precautions: In large amounts it is moderately toxic by swallowing. A mild allergen and skin irritant.

Synonyms: CAS: 103-41-3 ✦ BENZYL ALCOHOL CINNAMIC ESTER ✦ BENZYL γ-PHENYLACRYLATE ✦ CINNAMEIN ✦ trans-CINNAMIC ACID BENZYL ESTER ✦ 3-PHENYL-2-PROPENOIC ACID PHENYLMETHYL ESTER (9CI)

p-BENZYLPHENOL

Products and Uses: Commonly used in soaps and cleansers as an antiseptic and germicide.

Precautions: Toxic by swallowing. Possible allergen.

Synonyms: 4-HYDROXY DIPHENYLMETHANE

BENZYL SALICYLATE

Products and Uses: Ingredient used as perfume for cosmetics, musk perfume, sun-screening products, and soaps; also as a fixative, solvent, and odorant.

Precautions: A possible allergen. Skin irritant. Moderately toxic by swallowing.

Synonyms: CAS: 118-58-1 ✦ BENZYL-o-HYDROXY BENZOATE

BENZYL THIOL

Products and Uses: Usually used in beverages, baked goods, candy, and ice cream as an odorant and flavoring.

Precautions: Toxic by breathing and swallowing large amounts. Irritant to skin and eyes.

Synonyms: BENZYL MERCAPTAN ✦ α-TOLUENETHIOL

BERGAMOT OIL, rectified

Products and Uses: Obtained from the fruit of American horsemint plant. In bakery products, beverages (alcoholic), chewing gum, confections, gelatin desserts, ice cream, and puddings. In perfumery for hair tonic, oils, and dressings. An ingredient that affects taste or smell of product.

Precautions: Mildly toxic by swallowing large amounts. A mild skin irritant and allergen. GRAS (generally recognized as safe).

Synonyms: CAS: 8007-75-8 ✦ BERGAMOTTE OEL (GERMAN) ✦ OIL OF BERGAMOT, coldpressed ✦ OIL OF BERGAMOT, rectified

BETA CAROTENE

Products and Uses: Found in plants and raw fruits and vegetables. It is thought to be destroyed during cooking. It is used in orange beverages, desserts, cheese, ice cream, margarine, shortening, butter, nondairy whiteners, and cosmetics as a coloring and nutrient. Considered to be an anticancer compound.

Precautions: GRAS (generally recognized as safe) although excessive consumption can cause the skin to attain a yellow tinge, which is sometimes confused with jaundice. If the cause is beta carotene consumption, it is harmless.

Synonyms: VITAMIN A DERIVATIVE. ✦ β-CAROTENE

BHA and BHT

Products and Uses: Both are antioxidants commonly added in small amounts to food, particularly to oils and fats, as preservatives. Also used as preservative in rubber and plastics. Recent claims have been made that they inhibit cancers in lab animals, prevent herpes and other viruses, and increase life spans. There has been no scientific research to prove these theories.

Precautions: There has been some evidence that BHT could promote the growth of some cancers. There have also been allergic reactions reported by some.

Synonyms: BUTYLATED HYDROXYANISOLE ✦ BUTYLATED HYDROXYTOLUENE

BIOGAS

Products and Uses: Methane gas produced from animal manure by bacterial anaerobic digestion. Gas is then used as an energy source. Automobiles have been developed that use this fuel, for instance.

Precautions: Must be handled as cautiously as any gas, with an awareness of flammability and explosive potential.

Synonyms: MARSH GAS

BISMUTH

Products and Uses: Frequently used in pharmaceuticals, eye shadow, lipstick, bleaching skin creams, and hair dye products. This element is the basis for dozens of other chemicals used in medicines and cosmetics.

Precautions: A possible allergen. Poisonous to humans.

Synonyms: METALLIC ELEMENT OF ATOMIC NUMBER 83 ✦ BISMUTH-209

BISODIUM TARTRATE

Products and Uses: Found in fats, jams, jellies, margarine, meat products, oil, and sausage casings. As an emulsifier (stabilizes and maintains mixes), sequestrant (binds ingredients that affect the final product's appearance, flavor, or texture).

Precautions: Moderately toxic by swallowing gross amounts. GRAS (generally recognized as safe).

Synonyms: CAS: 868-18-8 ✦ DISODIUM TARTRATE ✦ DISODIUM *l*-(+)-TARTRATE ✦ SODIUM TARTRATE (FCC) ✦ SODIUM *l*-(+)-TARTRATE

BIS(TRIBUTYL TIN)OXIDE

Products and Uses: Applications include underwater paints, antifouling paints, and boat bottom paints. Used as fungicide and bactericide in paints.

Precautions: Toxic by swallowing and breathing. Moderately toxic by skin contact. A severe eye irritant. Caused animal birth defects.

Synonyms: CAS: 56-35-9 ✦ BIOMET TBTO ✦ BTO ✦ BUTINOX ✦ HEXABUTYLDITIN ✦ TRI-n-BUTYL-STANNAE OXIDE

BITHIONOL

Products and Uses: A component in deodorants, germicides, fungistats, and pharmaceuticals, as a preservative and bacteristat (kills bacteria).

Precautions: A skin irritant. FDA states it may not be used in cosmetics. EPA lists it on its Extremely Hazardous Substances list. Possible carcinogen (may cause cancer). A poison.

Synonyms: CAS: 97-18-7 ✦ ACTAMER ✦ BIDIPHEN ✦ BITHIONOL SULFIDE ✦ BITIN ✦ NEOPELLIS ✦ LOROTHIODOL

BLACK POWDER

Products and Uses: One of the oldest and most generally used explosives as blasting powder, fireworks, fuses, igniters, or gunpowder. In time fuses for blasting and shells. In igniter and primer assemblies for propellants, fireworks, and mining.

Precautions: Sensitive to heat. Does not detonate but is a dangerous fire and explosion hazard.

Synonyms: POTASSIUM NITRATE, CHARCOAL, SULFUR

BLEACH

Products and Uses: A 2 to 3% concentration of sodium hypochlorite. As clothing and textile bleach, algicide and disinfectant for pools, bactericide, deodorant, water purifier, and fungicide. For whitening, purifying, and disinfecting. In water treatment plants and industrial applications it is a 70% calcium hypochlorite solution.

Precautions: Possibly toxic. Fire risk when in contact with organic material.

Synonyms: HYDROGEN PEROXIDE ✦ SODIUM HYPOCHLORITE ✦ SODIUM PEROXIDE ✦ SODIUM CHLORITE ✦ CALCIUM HYPOCHLORITE ✦ HYPOCHLOROUS ACID ✦ SODIUM PERBORATE ✦ DI CHLORO DIMETHYL HYDANTOIN

BORAX PENTAHYDRATE

Products and Uses: A weed killer, soil sterilant, and fungus controller on citrus fruit. FDA limits residue on fruit. As a defoliant and fungicide.

Precautions: Toxic.

Synonyms: NONE FOUND.

BORIC ACID

Products and Uses: The mineral sassolite in the form of crystals, powder, or granules. A flame retardant, ingredient in glass and porcelain enameling, medical ointments, and eye washes. In coatings, for hobby work, and pharmaceuticals. A fungus preventer on citrus fruit.

Precautions: In concentrated amounts it is a poison by swallowing. Mildly toxic by skin contact. Swallowing may cause diarrhea, cramps, skin lesions, throat and mouth lesions, irregular heart beat, convulsions, and coma.

Synonyms: CAS: 10043-35-3 ✦ BORACIC ACID ✦ BOROFAX ✦ ORTHOBORIC ACID ✦ THREE ELEPHANT

BORIC OXIDE

Products and Uses: Various applications include paints and herbicides. Also used to add fire resistance to paints. It is a defoliant and pesticide.

Precautions: Moderately toxic by swallowing. An eye and skin irritant.

Synonyms: CAS: 1303-86-2 ✦ BORIC ANHYDRIDE ✦ BORON OXIDE

BOTULISM

Products and Uses: Spores of these bacteria are found virtually everywhere. The bacteria only produce toxins in an atmosphere of no oxygen and little acidity. Most potent bacterial poison known.

Precautions: Poison of toxin is produced by an organism called *Clostridium botulinum*. Found in soil and seawater. It can form spores that can contaminate food that is improperly processed, canned, preserved, or cooled. The spores can resist low cooking temperatures, then produce active organisms that in turn produce the toxins that taint the food. The toxin is one that affects the nerves of the peripheral nervous system. Double vision, dry mouth, nausea, vomiting, and progressive paralysis are symptoms. Humans and most animals are susceptible. Pigs, dogs, and cats are somewhat immune.

Synonyms: *CLOSTRIDIUM BOTULINUM*

BOVINE GROWTH HORMONE

Products and Uses: A hormone, abbreviated BST (bovine somatotropine) has now been thoroughly tested and approved by the FDA. Milk and meat are safe, they conclude. Cows produce about 10% more milk after treatment.

Precautions: According to the FDA, National Institute of Health, World Health Organization, Dept. of Health and Human Services, AMA, and so on, the milk and meat are safe to consume.

Synonyms: BGH ✦ BST ✦ BOVINE SOMATOTROPINE

BROMELIN

Products and Uses: Utilized in beer, bread, cereals (precooked), meat (raw cuts), poultry, and wine. For chillproofing of beer, milk-clotting enzyme, meat tenderizing, cereals (preparation of precooked), and tissue softening agent. Derived from pineapple juice.

Precautions: GRAS (generally recognized as safe).

Synonyms: CAS: 9001-00-7 ✦ ANANASE ✦ BROMELAINS ✦ EXTRANASE ✦ INFLAMEN ✦ PLANT PROTEASE CONCENTRATE ✦ TRAUMANASE

BROMINATED VEGETABLE (SOYBEAN) OIL

Products and Uses: Additive in soft drinks (fruit flavored). Also used as beverage stabilizer, flavoring agent, emulsifier, and clouding agent.

Precautions: FDA approves limited use.

Synonyms: VEGETABLE (SOYBEAN) OIL, BROMINATED ✦ BVO

BROMOACETONE

Products and Uses: Used for tear gas sprays, guns, aerosols, and bombs in crowd control and protection.

Precautions: Moderately toxic by breathing and skin contact. A lachrymator (causes eyes to water) and strong irritant.

Synonyms: CAS: 598-31-7 ✦ ACETONYL BROMIDE ✦ ACETYL METHYL BROMIDE ✦ BROMO-2-PROPANONE ✦ MONOBROMOACETONE

BROMOCHLOROMETHANE

Products and Uses: In fire extinguishers is a gas to extinguish flames.

Precautions: A poison. Mildly toxic by swallowing and breathing. Has a narcotic effect.

Synonyms: CAS: 74-97-5 ✦ METHYLENE CHLOROBROMIDE ✦ CHLOROBROMOMETHANE ✦ HALON 1011

BROMOTRIFLUOROMETHANE

Products and Uses: Utilized for refrigerant and fire extinguishment as a coolant and for fire protection.

Precautions: Toxic by breathing.

Synonyms: CAS: 75-63-8 ✦ TRIFLUOROBROMOMETHANE ✦ BROMOFLUOROFORM ✦ FREON 13B1 ✦ HALON 1301

BRONZE

Products and Uses: Applications include fine metallic flakes or powder in paint, powder cosmetics, hair coloring, pearl finish eye shadow, cosmetics, and decorative art materials.

Precautions: Powder is flammable.

Synonyms: COPPER/TIN

BUTANE

Products and Uses: Commonly used in cigarette, cigar, candle, outdoor grill, and fireplace lighters. The liquid fuel is under pressure and burns as a gas. An aerating agent, gas, and propellant.

Precautions: Mildly toxic via breathing. Causes drowsiness. An asphyxiant (causes suffocation). Very dangerous fire hazard when exposed to heat or flame. Highly explosive when exposed to flame. To fight fire, stop flow of gas. Narcotic in high concentrations. GRAS (generally recognized as safe).

Synonyms: CAS: 106-97-8 ✦ n-BUTANE ✦ BUTANEN ✦ DIETHYL ✦ METHYLETHYLMETHANE

BUTYL ACETATE

Products and Uses: A synthetic fruit flavoring for beverages, dessert ices, ice cream, candy, gum, baked goods, and gelatins. Used in perfumes, nail polishes, polish remover, and lacquers as a flavoring or solvent.

Precautions: Mildly toxic by breathing and swallowing large amounts. A skin and severe eye irritant. Effects by breathing are unspecified nasal and respiratory system problems. A mild allergen. High concentrations are irritating to eyes and respiratory tract and cause narcosis (semi-unconsciousness).

Synonyms: CAS: 123-86-4 ✦ ACETIC ACID n-BUTYL ESTER ✦ BUTILE ✦ n-BUTYL ACETATE ✦ 1-BUTYL ACETATE ✦ BUTYL ETHANOATE

tert-BUTYL ACETATE

Products and Uses: Gasoline additive and also used as a solvent.

Precautions: Flammable, moderate fire risk. Poison by breathing and swallowing.

Synonyms: CAS: 540-88-5 ✦ ACETIC ACID tert BUTYL ESTER ✦ TLA ✦ TEXACO LEAD APPRECIATOR

BUTYL ALCOHOL

Products and Uses: Varied uses include synthetic odorant in shampoos, solvent for polishes, and octane booster in gasoline. Also added to beverages, ice desserts, creams, candy, and baked goods. Frequently utilized in the production of fruit, liquor, butter flavorings, confectioneries, tablet form food supplements, and gum. Also used as color diluent and flavoring agent.

Precautions: Possible allergen. Moderately toxic by skin contact and swallowing. A severe eye irritation. Various respiratory system and nasal effects.

Synonyms: CAS: 71-36-3 ✦ 1-BUTANOL ✦ n-BUTANOL ✦ BUTAN-1-OL ✦ BUTANOL (DOT) ✦ n-BUTYL ALCOHOL ✦ BUTYL HYDROXIDE ✦ 1-HYDROXYBUTANE ✦ METHYLOLPROPANE ✦ PROPYLCARBINOL ✦ PROPYLMETHANOL

sec-BUTYL ALCOHOL

Products and Uses: Additive in paint remover and cleaners.

Precautions: Toxic on prolonged breathing and irritating to eyes and skin. Flammable, dangerous fire risk.

Synonyms: CAS: 78-92-2 ✦ SBA ✦ 2-BUTANOL ✦ METHYLETHYLCARBINOL

tert-BUTYL ALCOHOL

Products and Uses: An octane booster in unleaded gasoline (EPA approved). Used in perfumery and as a paint remover, solvent, gasoline additive, and odorant.

Precautions: Irritant to eyes and skin. Flammable, dangerous fire risk.

Synonyms: CAS: 75-65-0 ✦ 2-METHYL-2PROPANOL ✦ TRIMETHYL CARBINOL

BUTYL-p-AMINOBENZOATE

Products and Uses: A medicated ointment for burns with anesthetic properties. An ultraviolet (UV) absorber in suntan products; also a local anesthetic and a sunscreen.

Precautions: Toxic by swallowing.

Synonyms: CAS: 94-25-7 ✦ p-AMINOBENZOATE ACID BUTYL ESTER ✦ BUTAMBEN

BUTYL CELLOSOLVE

Products and Uses: Applications include haircoloring products; also as a dye or solvent.

Precautions: Poison by swallowing and skin contact. Moderately toxic via breathing. Can cause nausea, vomiting, nose tumors, headaches. Consumption caused animal birth defects.

Synonyms: CAS: 111-76-2 ✦ BUTOXYETHANOL ✦ BUTYL GLYCOL ✦ ETHYLENE GLYCOL MONOBUTYL ✦ POLY-SOLV EB

BUTYL ISOVALERATE

Products and Uses: A synthetic flavoring for beverages, ice desserts, creams, candy, bakery goods, and gelatins. In both chocolate and fruit flavorings.

Precautions: Mildly toxic by swallowing gross amounts.

Synonyms: CAS: 109-19-3 ✦ ISOVALERIC ACID, BUTYL ESTER ✦ n-BUTYL ISOPENTANOATE ✦ BUTYL-3-METHYLBUTYRATE

n-BUTYL LACTATE

Products and Uses: Applications include varnishes, inks, stencil pastes, perfumes, dry-cleaning fluids, and adhesives.

Precautions: Possible allergen. A skin irritant. Toxic when breathed.

Synonyms: CAS: 138-22-7 ✦ 2-HYDROXYPROPANIC ESTER ✦ LACTIC ACID BUTYL ESTER

BUTYL MERCAPTAN

Products and Uses: A warning odorant with a distinctive aroma, used in natural gas by power companies. Power companies use one drop for every 1000 cubic feet. The normal nose can smell less than 1%. At 7% the gas is explosive.

Precautions: Strong offensive odor. An eye irritant.

Synonyms: CAS: 109-79-5 ✦ BUTANETHIOL

BUTYL OLEATE

Products and Uses: Commonly used as a waterproofing agent in compounds, coatings, polishes, and for outdoor equipment. It is a solvent or lubricant.

Precautions: A possible allergen.

Synonyms: CAS: 142-77-8 ✦ WATER GUARD ✦ FAB GUARD

BUTYL PARABEN

Products and Uses: A preservative in a wide variety of cosmetic and food products. An antifungal preservative; inhibits mold growth.

Precautions: A skin irritant and possible allergen.

Synonyms: CAS: 94-26-8 ✦ BUTOBEN ✦ BUTYL CHEMOSEPT ✦ BUTYL PARASEPT ✦ BUTYL TEGOSEPT ✦ PARASEPT ✦ BUTYL-p-HYDROXYBENZOATE

BUTYRIC ACID

Products and Uses: An odorant in perfume ingredients, disinfectants, and gasoline; a flavoring in butter, candy, caramels, and fruit seasonings.

Precautions: Moderately toxic by swallowing large amounts. Strong irritant to skin and tissue. Possible allergen. Corrosive material.

Synonyms: CAS: 107-92-6 ✦ n-BUTYRIC ACID ✦ BUTANOIC ACID ✦ ETHYLACETIC ACID ✦ PROPYLFORMIC ACID

CABBAGE SEED OIL

Products and Uses: This oil is obtained from the *Brassica oleracea* plant for use in soaps and as a replacement olive oil.

Precautions: Considered safe when used under normal circumstances.

Synonyms: NONE KNOWN.

CACODYLIC ACID

Products and Uses: A defoliant in herbicide, soil sterilant, and for timber thinning.

Precautions: Toxic by swallowing. A skin and eye irritant. Possible carcinogen (may cause cancer).

Synonyms: CAS: 75-60-5 ✦ AGENT BLUE ✦ BOLLSEYE ✦ CHEXMATE ✦ DIMETHYL ARSENIC ACID ✦ ERASE ✦ SALVO

CADMIUM

Products and Uses: A heavy metal found in batteries, pigments, enamels, paints, photography, glazes, plating elements, plastic housewares, and current carriers.

Precautions: Flammable. Extremely toxic in repeated low concentrations. Can cause growth defects in children. A carcinogen (may cause cancer). Cadmium plating of food and beverage containers resulted in a number of outbreaks of gastroenteritis (food poisoning).

Synonyms: CAS: 7440-43-9 ✦ COLLOIDAL CADMIUM ✦ METALLIC ELEMENT OF ATOMIC NUMBER 48

CADMIUM TUNGSTATE

Products and Uses: An additive in fluorescent paint that produces crackle or crystal effects.

Precautions: Toxic by breathing.

Synonyms: DAYGLOW

CAFFEINE

Products and Uses: A stimulant in soft drinks (cola, orange), beverages (tea, coffee), and cocoa. It can still be found in blood 12 hours after drinking.

Precautions: GRAS (generally recognized as safe) by FDA when used at moderate levels. In excessive amounts it has been proven to be a human and animal poison by swallowing. However, it would take 40 cups of caffeinated coffee to cause death. Effects on the human body by swallowing include: ataxia (uncoordination), blood pressure elevation, convulsions, diarrhea, distorted perceptions, hallucinations, muscle contraction or spasticity, somnolence (general depressed activity), nausea or vomiting, tremors. A human teratogen causing developmental abnormalities of the craniofacial and musculoskeletal systems, miscarriage, or stillbirth. Human mutation (changes inherited characteristics) information reported. A possible carcinogen that caused cancer in animals. Large doses (above 1.0 g) cause palpitation, excitement, insomnia, dizziness, headache, and vomiting. Too frequent excessive use of caffeine in tea or coffee may lead to digestive disturbances, constipation, palpitations, shortness of breath, and depressed mental states. It is also implicated in cardiac disorders and fibrocystic breast problems. One cup of drip-brewed coffee has 115 mg. One cup of tea has 40 mg. A 12-oz cola drink has 46 mg. More than 300 mg a day can overstimulate the central nervous system.

Synonyms: CAS: 58-08-2 ✦ CAFFEIN ✦ COFFEINE ✦ ELDIATRIC C ✦ GUARANINE ✦ NO-DOZ ✦ ORGANEX ✦ THEIN

CALCIUM ACETATE

Products and Uses: Used in printing and dyeing, also in baked goods, cake mixes, fillings, gelatins, packaging materials, puddings, sweet sauces, syrups, toppings, and sausage casings. Its purpose is as a processing aid, sequestrant (affects the final products appearance, flavor, or texture), stabilizer, texturizing agent, thickening agent, and antimold agent.

Precautions: GRAS (generally recognized as safe) when used within limits specified by FDA.

Synonyms: CAS: 62-54-4 ✦ ACETATE of LIME ✦ BROWN ACETATE ✦ CALCIUM DIACETATE ✦ GRAY ACETATE ✦ LIME ACETATE ✦ LIME PYROLIGNITE ✦ SORBO-CALCIAN ✦ DIACETATE ✦ VINEGAR SALTS

CALCIUM ALGINATE

Products and Uses: Material derived from seaweed or kelp. In alcoholic beverages, baked goods, ice cream, confections, egg products, fats, frostings, canned fruits, gelatins, gravies, jams, oils, puddings, and sauces as an emulsifier, stabilizer, and thickening agent. Also used in waterproofing and medications.

Precautions: GRAS (generally recognized as safe).

Synonyms: CAS: 9005-35-0 ✦ ALGIN ✦ ALGINIC ACID

CALCIUM BENZOATE

Products and Uses: A preservative in margarine.

Precautions: GRAS (generally recognized as safe).

Synonyms: NONE KNOWN.

CALCIUM BROMATE

Products and Uses: Commonly used in baked goods as a dough conditioner and maturing agent.

Precautions: GRAS (generally recognized as safe).

Synonyms: NONE FOUND.

CALCIUM CARBONATE

Products and Uses: Found in toothpastes and as a filler in deodorants and cosmetics. In baking powder, chewing gum, desserts (dry mix), dough, packaging materials, and wine. It is used as an antacid, an antidiarrheal, an alkali, dietary supplement, dough conditioner, firming agent, modifier for chewing gum, nutrient, release agent for chewing gum, stabilizer, texturizing agent, chewing gum texturizer, and yeast food.

Precautions: A severe eye and moderate skin irritant. GRAS (generally recognized as safe). A common air contaminant. It is considered the most concentrated and cheapest form of commercial calcium supplement. It is best absorbed when taken with food. The recommended daily allowance is about 800 mg.

Synonyms: CAS: 1317-65-3 ✦ AGRICULTURAL LIMESTONE ✦ AGSTONE ✦ ARAGONITE ✦ ATOMIT ✦ CALCITE ✦ CHALK ✦ DOLOMITE ✦ LIMESTONE ✦ LITHOGRAPHIC STONE ✦ MARBLE ✦ PORTLAND STONE

CALCIUM CHLORIDE

Products and Uses: A deicer for winter roads. Used in apple slices, baked goods, beverage bases (nonalcoholic), beverages (nonalcoholic), cheese, coffee, condiments, dairy product analogs, fruits (processed), fruit juices, gravies, jams (commercial), jellies (commercial), meat (raw cuts), meat products, milk (evaporated), pickles, plant protein products, potatoes (canned), poultry (raw cuts), relishes, sauces, tea, tomatoes (canned), and vegetable juices (processed). As an anticaking agent, antimicrobial agent, curing agent, firming agent, flavor enhancer, humectant (moisturizer), nutrient supplement, pickling agent, sequestrant (affects the final product's appearance flavor or texture), stabilizer, texturizing agent, and thickening agent. Found in pharmaceuticals as an antiseptic or diuretic.

Precautions: In large amounts it is moderately toxic by swallowing. Possible carcinogen (may cause cancer) that caused tumors in animal studies. FDA states GRAS (generally recognized as safe) when used within limits.

Synonyms: CAS: 10043-52-4 ✦ CALPLUS ✦ CALTAC ✦ DOWFLAKE ✦ LIQUIDOW ✦ PELADOW ✦ SNOMELT ✦ SUPERFLAKE ANHYDROUS

CALCIUM CITRATE

Products and Uses: A dietary supplement, firming agent and nutrient. In beans (lima), flour, and peppers. As a buffer (regulates the acidity or alkalinity).

Precautions: GRAS (generally recognized as safe).

Synonyms: CAS: 813-94-5 ✦ LIME CITRATE ✦ TRICALCIUM CITRATE

CALCIUM DISODIUM EDTA

Products and Uses: Commonly indicated as a preservative and sequestrant (binds constituents that affect the final product's appearance, flavor or texture). In beverages (distilled alcoholic), beverages (fermented malt), cabbage (pickled), clams (cooked canned), corn (canned), crabmeat (cooked canned), cucumbers (pickled), dressings (nonstandardized), egg product (that is hard-cooked and consists, in a cylindrical shape, of egg white with an inner core of egg yoke), French dressing, lima beans (dried, cooked canned), margarine, mayonnaise, mushrooms, pecan pie filling, pinto beans (processed dry), potato salad, potatoes (canned white), salad dressings, sauces, shrimp (cooked canned), soft drinks (canned carbonated), spreads (sandwich), and spreads (artificially colored and lemon flavored or orange flavored).

Precautions: FDA states GRAS (generally recognized as safe) when used within stated limits.

Synonyms: CALCIUM DISODIUM EDETATE ✦ CALCIUM DISODIUM ETHYLENEDIAMINETETRAACETATE ✦ CALCIUM DISODIUM (ETHYLENEDINITRILO)TETRAACETATE

CALCIUM GLUCONATE

Products and Uses: Found in vitamin tablets, baked goods, dairy product analogs (simulations), gelatins, gels, puddings, and sugar substitutes. As a firming agent, formulation aid, sequestrant (binds constituents that affect the final product's appearance, flavor or texture), stabilizer (used to keep a uniform consistency), texturizing agent, and thickening agent.

Precautions: Possible allergen. GRAS (generally recognized as safe) when used within FDA limitations.

Synonyms: CAS: 299-28-5

CALCIUM HYDROXIDE

Products and Uses: Various applications include brick mortar, plasters, cements, depilatory (hair remover), disinfectants, water softener, and purifier of sugar juices. Also useful as food additive buffer, firming agent, neutralizing agent, and miscellaneous general purpose food chemicals. Also an important agricultural commodity.

Precautions: Mildly toxic by swallowing large amounts. A severe eye irritant. A skin, nose, throat, and respiratory system irritant. Causes dermatitis (skin rash or irritation). A common air contaminant. GRAS (generally recognized as safe) when used at moderate levels to accomplish the intended results.

Synonyms: CAS: 1305-62-0 ✦ BELL MINE ✦ CALCIUM HYDRATE ✦ HYDRATED LIME ✦ KEMIKAL ✦ LIME WATER ✦ SLAKED LIME

CALCIUM HYPOCHLORITE

Products and Uses: A chemical compound containing chlorine. An algicide and bactericide for swimming pools. A deodorant, water purifier, fungicide, and bleaching agent. Used as a disinfectant and whitener.

Precautions: Moderately toxic by swallowing. Can cause severe irritation of nose, throat, and skin. Can cause fumes capable of producing pulmonary edema. A mutagen (changes inherited characteristics). Dangerous fire risk.

Synonyms: CAS: 7778-54-3 ✦ CALCIUM OXYCHLORIDE ✦ BLEACHING POWDER ✦ CALCIUM HYPOCHLORIDE ✦ HTH ✦ HY-CHLOR ✦ LIME CHLORIDE ✦ PERCHLORON

CALCIUM IODATE

Products and Uses: Common additive in bread, flour, deodorant, and mouth wash, as a dough conditioner, maturing agent, food additive, and deodorizer.

Precautions: A nuisance dust in dry form. GRAS (generally recognized as safe) by FDA when used within limits. Fire risk when in contact with organic materials.

Synonyms: CAS: 7789-80-2

CALCIUM LACTATE

Products and Uses: Used in bread, cake (angel food), fruits (canned), meat food sticks, meringues, milk (dry powder), sausage, sausage (imitation), vegetables (canned), and whipped toppings. As a preservative, buffer, dough conditioner, firming agent, flavor enhancer, flavoring agent, leavening agent, nutrient supplement, stabilizer, thickening agent, additive, and yeast food.

Precautions: GRAS (generally recognized as safe) by FDA when used within limits stated except for infant foods and formulas.

Synonyms: CAS: 814-80-2

CALCIUM LACTOBIONATE

Products and Uses: Found in pudding mixes (dry) as a firming agent.

Precautions: Harmless when used for intended purposes.

Synonyms: CAS: 5001-51-4 ✦ CALCIUM 4-(β-d-GALACTOSIDO)-d-GLUCONATE

CALCIUM NITRATE

Products and Uses: Used in fireworks, explosives, matches, and fertilizers.

Precautions: Forms powerfully explosive mixtures with aluminum + ammonium nitrate + formamide + water, ammonium nitrate + hydrocarbon oils, ammonium nitrate + water soluble fuels and organic materials. May cause irritation of throat and nasal passages, causes eye tearing and pain. Overexposure could cause nausea, head pains, vomiting, shortness of breath, coma, and death.

Synonyms: CAS: 10124-37-5 ✦ LIME SALTPETER ✦ NITROCALCITE

CALCIUM OXIDE

Products and Uses: Widely used in the food industry. Also in building products, sewage treatment, pulp and paper industry, and in insecticides. In poultry feeds, sugar, insecticides, and fungicides. Sugar refining agent, food additive, dietary supplement, dough conditioner, nutrient, and yeast food. Before electricity, an oxyhydrogen flame impinged on a cylinder of lime, which caused a brilliant white light that was concentrated to a beam by a lens. The light was used as a spotlight for stage shows, thus the phrase, "in the limelight."

Precautions: A common air contaminant. Very irritating to skin, eyes, nose and throat. At full strength, may cause difficult breathing and pulmonary edema. Upon ingestion may cause hemorrhaging, perforation of the esophagus and stomach, and death.

Synonyms: CAS: 1305-78-8 ✦ BURNT LIME ✦ CALCIA ✦ CALX ✦ LIME ✦ QUICKLIME

CALCIUM PEROXIDE

Products and Uses: Various applications include bakery products, seed disinfectants, tooth powders and pastes. Also used as a dough conditioner, oxidizing agent, and antiseptic.

Precautions: Irritating in concentrated form. Will react with moisture to form slaked lime. A strong alkali. An oxidizer. GRAS (generally recognized as safe) when used in moderate amounts.

Synonyms: CAS: 1305-79-9 ✦ CALCIUM DIOXIDE ✦ CALCIUM SUPEROXIDE

CALCIUM PHOSPHIDE

Products and Uses: Component in signal fires, rat poisons, fireworks, and tor-pedoes.

Precautions: Dangerous fire risk. Highly toxic.

Synonyms: PHOTOPHOR ✦ DIPHOSPHIDE ✦ TRICALCIUM

CALCIUM PHYTATE

Products and Uses: Derived from corn steep liquor. Used as a sequestering agent (affects the final appearance, flavor or texture of product) to remove metals from wine and vinegar, also in pharmaceuticals and nutrients, as a binding agent, a calcium supplement, and diet supplement.

Precautions: Harmless when used for intended purpose.

Synonyms: HEXACALCIUM PHYTATE

CALCIUM PROPIONATE

Products and Uses: Found in cheese, confections, dough (fresh pie), fillings, and frostings, gelatins, jams, jellies, pizza crust, puddings, tobacco, and pharmaceu-ticals. Used in food and cosmetics as an antimicrobial agent, mold inhibitor, preservative, additive, and antifungal.

Precautions: GRAS (generally recognized as safe).

Synonyms: CAS: 4075-81-4

CALCIUM PYROPHOSPHATE

Products and Uses: An additive used as a buffer, dietary supplement, neutralizing agent, and nutrient. For polishing agents in toothpaste or powder.

Precautions: GRAS (generally recognized as safe).

Synonyms: CAS: 7790-76-3

CALCIUM RESINATE

Products and Uses: Various uses include waterproofers, paint driers, perfumes, cosmetics, enamels, and soaps. Also useful as a coating for fabrics, gel thick-ener, and detergents.

Precautions: Flammable. Dangerous fire risk.

Synonyms: CAS: 9007-13-0 ✦ LIMED ROSIN

CALCIUM SORBATE

Products and Uses: Found in syrups (chocolate and fruit), fresh fruit salad, beverages, bakery goods, cheesecake, cheese, jellies, and salads (slaw, gelatin, macaroni, potato). Used as a mold retardant and preservative.

Precautions: FDA states GRAS (generally recognized as safe) when used at moderate levels to accomplish the desired results.

Synonyms: NONE FOUND.

CALCIUM STEARATE

Products and Uses: An additive in beet sugar, candy (pressed), garlic salt, meat tenderizer, molasses, salad dressing mix (dry), vanilla, yeast, hair products, and paints. As as anticaking agent, binder, emulsifier, flavoring agent, lubricant, release agent, stabilizer, thickening agent, and coloring agent.

Precautions: FDA approves use at moderate levels to accomplish the desired results.

Synonyms: CAS: 1592-23-0

CALCIUM STEAROYL LACTATE

Products and Uses: Usually in bakery products (yeast leavened), coffee whiteners, egg white (dried), egg white (liquid and frozen), margarine (low fat), potatoes (dehydrated), puddings, and artificial whipped cream. Useful as a dough conditioner, stabilizer, and whipping agent.

Precautions: FDA approves limited use. Must conform to limitations.

Synonyms: CALCIUM STEAROYL-2-LACTATE

CALCIUM SULFIDE

Products and Uses: Various applications include as luminous paint, depilatories, acne medication, oil additive; also as a preservative in paints and in skin products.

Precautions: Poison via breathing. Possible allergen. Irritating to skin, nose and throat.

Synonyms: CAS: 20548-54-3 ✦ CALCIC LIVER OF SULFUR ✦ HEPAR CALCIS ✦ OLDHAMITE ✦ CALCIUM SULPHIDE

CALCIUM SULFITE

Products and Uses: Frequently used as a sugar and brewing disinfectant. For cleansing and preserving; it prevents discoloring.

Precautions: Harmless when used for intended purpose.

Synonyms: NONE FOUND.

CALCIUM THIOGLYCOLLATE

Products and Uses: Used in depilatory and hair-waving products. For hair removal and hair permanents.

Precautions: Possible allergen. Possible skin irritant.

Synonyms: NONE FOUND.

CAMPHOR

Products and Uses: Utilized in skin lotion, moth and mildew proofings, tooth powders, lacquers, and insecticides, also for fragrances, flavorings, and conditioners.

Precautions: Toxic by breathing and swallowing and absorption through skin. Skin irritant. Swallowing causes nausea, vomiting, dizziness, and convulsions.

Synonyms: CAS: 76-22-2 ✦ GUM CAMPHOR ✦ 2-CAMPHANONE

CAMPYLOBACTERIOSIS

Products and Uses: A bacteria found in raw poultry, meat, and unpasteurized milk.

Precautions: A food-borne illness that can cause diarrhea, abdominal cramps, fever, and possibly bloody stools. Conditions can last for a week or more.

Synonyms: *CAMPYLOBACTER JEJUNI* ✦ ROD SHAPED BACTERIA

CANANGA OIL

Products and Uses: A spice and food flavoring agent used in carbonated drinks (cola, ginger ale), ice desserts, candies, and bakery goods.

Precautions: Possible allergen. GRAS (generally recognized as safe) when used in moderate amounts.

Synonyms: NONE FOUND.

CANDELILLA WAX

Products and Uses: Frequently used for lubricants, masticatory substance in chewing gum bases, surface-finishing agents, waxing polishes, and watersealing. In candy (hard), chewing gum, leather dressings, polishes, cements, varnishes, candles, sealing wax, waterproofing, insect repellent, paint remover, dental waxes, and food coatings.

Precautions: FDA states GRAS (generally recognized as safe) when used in moderate amounts.

Synonyms: CAS: 8006-44-8

CANTHAXANTHIN

Products and Uses: Utilized in carbonated sodas, feeds (broiler chicken), salad dressings, spaghetti sauce, and artificial tanning products. As color additive (reddish) for food and drugs.

Precautions: Swallowing may cause night blindness. FDA permits limited use in foods.

Synonyms: CAS: 514-78-3 ✦ CANTHA ✦ β-CAROTENE-4,4′-DIONE ✦ 4,4′-DIKETO-β-CAROTENE

CAPSAICIN

Products and Uses: Currently popular chemical in topical arthritis medications, spicy seasonings, and self-defense sprays. This is the substance that makes chili peppers hot.

Precautions: Irritating and burning sensation to nose and throat that cannot by alleviated by water. However, milk, sour cream or yogurt has cooling effect.

Synonyms: HOT PEPPERS ✦ TABASCO

CAPTAN

Products and Uses: Used as a seed treatment, fungicide, preservative, and bacteriostat in almonds, animal feed, apples, beans, beef, beets, broccoli, cabbage, carrots, corn, garlic, kale, lettuce, peaches, peas, pork, potatoes, raisins, spinach, and strawberries. Also in paints, plastics, leather, and fabrics.

Precautions: Toxic by inhalation and an irritant to skin and mucous membrane. Produce should be washed well before consuming!

Synonyms: CAS: 133-06-2 ✦ AGROSOL S ✦ AMERCIDE ✦ BANGTON ✦ CAPTANE ✦ CAPTAN-STREPTOMYCIN 7.5-0.1 POTATO SEED PIECE PROTECTANT ✦ CAPTEX ✦ FLIT 406 ✦ HEXACAP ✦ KAPTAN ✦ MERPAN ✦ ORTHOCIDE ✦ OSOCIDE ✦ VANICIDE

CARBANOLATE

Products and Uses: Used as insecticide and nematocide (kills parasitic worms) on animal feed, bananas, beans, citrus fruit, coffee, hops (dried), peanuts, pecans, potatoes, sorghum, soybeans, sugar beets, sugarcane, and sweet potatoes.

Precautions: Deadly poison by swallowing or skin contact. Human mutagen (changes inherited characteristics). In 1985 over 150 people in California exhibited toxic effects from eating watermelons contaminated with aldicarb. FDA permits limited use. On EPA Extremely Hazardous Substances list.

Synonyms: CAS: 116-06-3 ✦ ALDECARB ✦ ALDICARB ✦ AMBUSH ✦ TEMIC ✦ TEMIK

CARBARYL

Products and Uses: An insecticide commonly used for pet flea powder preparations. It is banned in Germany because of causing lab animal mutations.

Precautions: Poison by ingestion. Can cause severe eye and skin irritation. Highly toxic and can be absorbed through the skin.

Synonyms: CAS: 63-25-2 ✦ SEVIN ✦ METHYL-CARBARATE-1-NAPTHOLENOL

CARBOLIC ACID; see PHENOL

CARBON BLACK

Products and Uses: A color additive used in printing inks, carbon paper, typewriter ribbons, and paint pigments.

Precautions: Mildly toxic by swallowing, breathing, and skin contact. Possible carcinogen (may cause cancer). A nuisance dust in high concentrations. While it is true that the tiny particulates of carbon black contain some molecules of carcinogenic materials, the carcinogens are apparently held tightly and are not released by hot or cold water, gastric juices, or blood plasma. FDA no longer permits use in food.

Synonyms: CAS: 1333-86-4 ✦ ACETYLENE BLACK ✦ CHANNEL BLACK ✦ FURNACE BLACK ✦ LAMP BLACK

CARBON DIOXIDE

Products and Uses: Used as aerating agent, for carbonation, cooling agent, gas, leavening agent, modified atmospheres for pest control, dry ice, propellant, fire extinguishers, and enrichment of greenhouse air. Found in beverages (carbonated), fruit, meat, poultry, and wine.

Precautions: An asphyxiant. Contact of carbon dioxide snow with the skin can cause burns. FDA states GRAS (generally recognized as safe) when used at moderate levels to accomplish the desired results.

Synonyms: CAS: 124-38-9 ✦ CARBONIC ACID GAS ✦ CARBONIC ANHYDRIDE

CARBON MONOXIDE

Products and Uses: A gas that is colorless, odorless, and tasteless, resulting from the incomplete burning of materials. Produced by faulty gas furnaces and other gas appliances. Also present in coal, tobacco, and wood smoke.

Precautions: A deadly, odorless, poisonous gas, which causes indoor pollution problems. Even at low levels it can cause health risks. Symptoms can include drowsiness, headaches, and dizziness. At high levels it can cause suffocation. Those with heart, lung, and blood disorders are most vulnerable.

Synonyms: CAS: 630-08-0 ✦ CARBONIC OXIDE ✦ EXHAUST GAS ✦ FLUE GAS ✦ KOHLENKOYD (GERMAN)

CARBON TETRACHLORIDE

Products and Uses: Banned from all uses in 1996 due to ozone depletion concerns. Previously found in spot removers, fire extinguishers and dry-cleaning products as a fumigant, solvent, and pesticide.

Precautions: A poison by swallowing. Mildly poisonous by breathing. An eye and skin irritant. Damages liver, kidneys, and lungs. It has a narcotic action resembling chloroform. Probably a carcinogen (causes cancer).

Synonyms: CAS: 56-23-5 ✦ BENZINOFORM ✦ CARBON CHLORIDE ✦ CARBON TET ✦ METHANE ✦ NECATORINE ✦ TETRAFORM ✦ TETRASOL ✦ METHANE TETRACHLORIDE

CARBOXYMETHYLCELLULOSE

Products and Uses: A thickening agent in detergents, soaps, dietetic foods, ice cream, emulsion paints, cosmetics (toothpaste, hair products, blusher, shaving creams), laxatives, and antacids as an emulsifier, stabilizer, and foaming agent.

Precautions: In excessive amounts it is mildly toxic by swallowing. It caused animal reproductive effects (infertility, or sterility, or birth defects). Possible carcinogen (may cause cancer). It migrates to food from packaging materials.

Synonyms: CAS: 9004-32-4 ✦ CARMETHOSE ✦ CELLUGEL ✦ AQUAPLAST ✦ CELLPRO

CARMINE

Products and Uses: A red color additive in dyes, inks, and food color.

Precautions: Derived from insects; therefore, it must be pasteurized to destroy salmonella. A possible allergen.

Synonyms: CAS: 1390-65-4 ✦ B ROSE LIQUID ✦ CARMINIC ACID

CARNAUBA WAX

Products and Uses: A natural product derived from palm leaves. Found in baked goods, baking mixes, candy (soft), chewing gum, confections, frostings, fruit juices, fruit juices (processed), fruits (fresh), fruits (processed), gravies, and sauces. Used as a texturizer, an anticaking agent, candy glaze, candy polish, formulation aid, lubricant, release agent, in makeup base, blusher, mascara, lipstick, dipilatory (hair remover), antiperspirant, and liquid makeup.

Precautions: FDA states GRAS (generally recognized as safe).

Synonyms: CAS: 8015-86-9 ✦ BRAZIL WAX

CARRAGEEN

Products and Uses: Derived from dried seaweed. In dairy products, dessert gels (water), jelly (low calorie), meat (restructured), poultry, cosmetic oil, chocolate, toothpaste, ice cream, and chocolate milk. As a binder, emulsifier, extender, gelling agent, stabilizer, and thickening agent.

Precautions: GRAS (generally recognized as safe) by FDA when used within limits. A suspected carcinogen (may cause cancer) that caused tumors in animals.

Synonyms: CAS: 9000-07-1 ✦ AUBYGEL ✦ AUBYGUM ✦ CARASTAY ✦ CARRAGEENAN GUM ✦ GALOZONE ✦ GELCARIN ✦ GELOZONE ✦ GENU ✦ GENUGEL ✦ IRISH GUM ✦ IRISH MOSS EXTRACT ✦ IRISH MOSS GELOSE

CARVACROL

Products and Uses: An additive in perfume, fungicides, disinfectants, citrus fruit, mint, and spice flavorings for beverages, ice desserts, candies, bakery goods, and spicy condiments. A natural oil that has been synthesized. Also used as a flavoring and as a germicide.

Precautions: Poison by swallowing in large amounts. Moderately toxic by skin contact. A severe skin irritant. Combustible liquid. FDA approves use at moderate levels to accomplish the desired results.

Synonyms: CAS: 499-75-2 ✦ 2-p-CYMENOL ✦ ISOPROPYL-o-CRESOL ✦ 5-ISOPROPYL-2-METHYLPHENOL ✦ ISOTHYMOL ✦ 2-METHYL-5-ISOPROPYLPHENOL ✦ o-THYMOL

d-CARVONE

Products and Uses: Additive in bakery products, beverages (alcoholic), beverages (nonalcoholic), chewing gum, condiments, confections, ice cream, liquors, perfumery, and soaps. Useful as a flavoring and perfuming agent. This (*d*-) form is the main constituent of caraway and dill oils.

Precautions: In large amounts it is a poison by swallowing and skin contact. A skin irritant. FDA approves use at moderate levels to accomplish the desired results.

Synonyms: CAS: 2244-16-8 ✦ (+)-CARVONE ✦ *d*(+)-CARVONE ✦ (S)-CARVONE ✦ (S)-(+)-CARVONE

l-CARVONE

Products and Uses: Frequently used in beverages, ice desserts, candies, gum, bakery goods, perfumes, toilet soaps, OTC (over-the-counter) medications, candy mints, and breath mints. Flavoring for liqueur, mints, and spices. This (*l*-) form occurs principally in spearmint oil.

Precautions: Moderately toxic by swallowing gross amounts. FDA approves use at moderate levels to accomplish the desired results.

Synonyms: CAS: 6485-40-1 ✦ (-)-CARVONE ✦ (R)-CARVONE

CASCARILLA OIL

Products and Uses: A flavoring agent derived from a tree bark grown in the Caribbean, used in smoking tobacco and fruit and soft drinks as well as ice cream, candy, and baked products. Produces a spicy odor and taste.

Precautions: Generally considered harmless.

Synonyms: SWEETWOOD BARK OIL

CASEIN

Products and Uses: A nutritious milk protein containing all the essential amino acids. Used in a wide variety of products including cheese, interior paints, adhesives for laminates (plywood), feeds, and dietetic preparations. Acts as a thickener, whitening agent, binder, coating, glue, and sizing.

Precautions: Probably harmless when used for intended purposes.

Synonyms: CAS: 9005-46-3 ✦ SODIUM CASEINATE ✦ CASEIN AND CASEINATE ✦ NUTROSE

CASHEW GUM

Products and Uses: Various uses include inks, insecticides, glues, tanning agents, varnishes, and bookbinders' gum. Binder derived from exudate from cashew nut tree.

Precautions: Possible allergen. Harmless when used for intended purposes.

Synonyms: ANACARDIUM GUM

CASSIA OIL

Products and Uses: A flavoring in perfumes, medications, foods, beverages, odor-ants, soaps, and laxatives. Similar to a clove/cinnamon aroma.

Precautions: Possible allergen. Poison by skin contact. Moderately toxic by swal-lowing. A human skin irritant. GRAS (generally recognized as safe) when used in moderate amounts.

Synonyms: CAS: 8007-80-5 ✦ ARTIFICIAL CINNAMON OIL ✦ CINNAMON BARK OIL ✦ CINNAMON BARK OIL, CEYLON TYPE (FCC) ✦ CINNAMON OIL

CASTOR OIL

Products and Uses: A nonvolatile oil from the seeds of the castor bean. Used in candy (hard), vitamin and mineral tablets, lipstick, laxatives, bath oils, soaps, hair products, solid perfumes, nail polish, and nail polish remover. Also used as an antisticking agent and in coatings and medications.

Precautions: Moderately toxic by swallowing large amounts. An allergen. An eye irritant. Combustible when exposed to heat. Spontaneous heating may occur. Use at moderate levels to accomplish the desired results.

Synonyms: CAS: 8001-79-4 ✦ AROMATIC CASTOR OIL ✦ CASTOR OIL, AROMATIC ✦ COSMETOL ✦ CRYSTAL O ✦ GOLD BOND ✦ OIL OF PALMA CHRISTI ✦ PHORBYOL ✦ RICINUS OIL ✦ TANGANTANGAN OIL ✦ TURKEY RED OIL

CATIONIC IMIDAZOLINES

Products and Uses: Found in cosmetics and cleaners as an emulsifier (stabilizes and maintains mixes to aid in suspension of oily liquids), antistatic agent, and water displacer.

Precautions: Could cause allergic symptoms in susceptible individuals.

Synonyms: MONAZOLINES

CELLOPHANE

Products and Uses: A natural, nonsynthetic, commonly used packaging material. It is produced from wood pulp, or cellulose, by the viscose process.

Precautions: It does not melt, but burns readily and is not self-distinguishing. It is biodegradable.

Synonyms: NONE FOUND.

CELLULOSE

Products and Uses: The fibrous material in the cell walls of trees and plants. It is used to make rayon, natural cellulose sponges, and building insulation. Basis for wood, paper, linen, and cotton.

Precautions: Harmless when used appropriately.

Synonyms: NONE FOUND.

CETYL ALCOHOL

Products and Uses: A nondrying emollient, conditioner or softener used in facial makeup, hair products, blushes, mascara, lipstick, nail polish remover, deodorant, and baby skin products. Also used in foods. A solvent.

Precautions: Relatively nontoxic by swallowing. Possible allergen.

Synonyms: CAS: 36653-82-4 ✦ 1-HEXADECANOL ✦ ADOL ✦ ALCOHOL C-16 ✦ CETYLOL ✦ ETHAL ✦ ETHOL ✦ PALMITYL ALCOHOL

CETYL PYRIDINIUM CHLORIDE

Products and Uses: In cough lozenges, and cough syrups, it is used as an antibacterial.

Precautions: Poison by swallowing gross amounts. Moderately toxic by skin contact. A skin and eye irritant.

Synonyms: CAS: 123-03-5 ✦ CEEPRYN ✦ CEPACOL ✦ CEPACOL CHLORIDE ✦ CETAMIUM ✦ DOBENDAN ✦ MEDILAVE ✦ PRISTACIN ✦ PYRISEPT

CFC; see CHLOROFLUOROCARBON

CHAMOMILE OIL

Products and Uses: Found in hair products, conditioners, rinses, and skin products. Considered by herbalists to be an anti-inflammatory for skin and mucous membrane problems, medication for indigestion, menstrual cramps, muscle spasms, and so on. The oil is distilled from tiny flowers. Some varieties are used in food and beverages for flavorings of vanilla, maple, berry, or spice.

Precautions: People who are allergic to ragweed or flowers in the daisy family could suffer allergic reactions. GRAS (generally recognized as safe) when used in moderate amounts.

Synonyms: CAS: 8002-66-2 ✦ CAMOMILE OIL GERMAN ✦ CHAMOMILE-GERMAN OIL ✦ GERMAN CHAMOMILE OIL ✦ HUNGARIAN CHAMOMILE OIL

CHAPARRAL

Products and Uses: A herb considered to be harmful by the U.S. FDA. It is sold as a tea, tablet, and capsule. Sometimes promoted as a cure for everything from acne to cancer. Frequently it is in combination herbal formulations.

Precautions: Caused documented cases of acute nonviral hepatitis (rapidly developing liver damage).

Synonyms: VARIOUS BRAND NAMES.

CHARCOAL

Products and Uses: A porous form of carbon, used as a decolorizing agent, odor-removing agent, purification agent in fat and beverage processing, and as a taste-removing agent. Wood charcoal is used as fuel. Coconut shell charcol is used as a gas absorbant. Animal charcoal, produced by heating bones and dissolving out the calcium phosphates and mineral salts with acid, is used in sugar refining.

Precautions: It can cause a dust irritation, particularly to the eyes, nose, and throat. Combustible when exposed to heat. Dust is flammable and explosive when exposed to heat or flame. GRAS (generally recognized as safe) when used within limitations.

Synonyms: CAS: 64365-11-3 ✦ ACTIVATED CARBON ✦ CARBON, ACTIVATED ✦ AMORPHOUS CARBON

CHICLE

Products and Uses: Ingredient in chewing gum, acts as a softener. Derived from latex of sapodilla tree in Mexico and Central America.

Precautions: Swallowing should be avoided.

Synonyms: THERMOPLASTIC SUBSTANCE

CHLORAL HYDRATE

Products and Uses: Found in sedative, narcotic, anticonvulsant, and anesthetic medications. Used externally as a liniment.

Precautions: Overdose is toxic. Hypnotic drug. Damaging to eyes.

Synonyms: CAS: 302-17-0 ✦ KNOCKOUT DROPS ✦ TRICHLOR ACETALDEHYDE ✦ HYDRATED TRICHLOROETHYLIDENE GLYCOL

CHLORBENSIDE

Products and Uses: An acaricide. Kills mange mites.

Precautions: Toxic by swallowing. Possibly a skin irritant.

Synonyms: CAS: 103-17-3 ✦ CHLOROCIDE ✦ CHLOROPARACIDE ✦ CHLOROBENZIDE ✦ CHLOROPARACIDE

CHLORDANE

Products and Uses: Insecticide. Used as a pesticide and fumigant. It has been unavailable for residential use in the U.S. since 1987.

Precautions: Suspected carcinogen that caused cancer in animals. Poison to humans by swallowing. Can be absorbed through the skin, resulting in toxic effects. Swallowing or skin contact results in tremors, convulsions, excitement, ataxia (loss of muscle coordination), and gastritis. Combustible liquid. It is no longer permitted for use as a termiticide in homes.

Synonyms: CAS: 57-74-9 ✦ ASPON-CHLORDANE ✦ BELT ✦ CHLORDAN ✦ γ-CHLORDAN ✦ CHLORINDAN ✦ CHLOR KIL ✦ CHLORODANE ✦ CHLORTOX ✦ TOXICHLOR

CHLORINE

Products and Uses: Extremely useful household chemical. Used for laundry bleach, bath cleansers, tile cleansers, and cleaning liquids. As an antimicrobial agent, bleaching agent, oxidizing agent. Chlorine is used industrially to bleach paper products; this process results in highly toxic compounds known as dioxins.

Precautions: Moderately toxic by breathing. A human mutagen (changes inherited characteristics). Some respiratory system effects by breathing are changes in the trachea or bronchi, emphysema, chronic pulmonary edema or congestion. A strong irritant to eyes, nose, and throat.

Synonyms: CAS: 7782-50-5 ✦ BERTHOLITE ✦ CHLORINE MOL. ✦ MOLECULAR CHLORINE

CHLOROACETOPHENONE; see MACE

CHLOROFLUOROCARBON

Products and Uses: Compounds that consist of carbon, chlorine, and fluorine. Used as coolant for refrigerators and air conditioners. Also used in aerosol propellants, blowing agents, and solvents.

Precautions: High concentrations cause narcosis and anesthesia in humans. Harmful effects are eye irritation and liver changes. Poison by breathing. Exposure could cause blindness. It is believed to be depleting stratospheric (upper atmosphere) ozone layer, thus causing global warming. Ozone layer depletion has also resulted in an increase in skin cancers and cataracts. In the next few years all products containing CFCs will have been reformulated and CFC use will be prohibited. Many countries agreed in 1987 to phase out chlorofluorocarbons. China agreed to begin their ban in 1998.

Synonyms: CAS: 75-69-4 ✦ CFC ✦ TRICHLORO MONOFLUORO METHANE ✦ FREON 11 ✦ ARCTON 9 ✦ FRIGEN 11 ✦ HALOCARBON 11

CHLORO-IPC

Products and Uses: A herbicide. A preemergent sprayed on potatoes to prevent sprouting, therefore providing longer shelf life.

Precautions: Hazard by swallowing. Possible carcinogen (may cause cancer). A human mutagen (changes inherited characteristics).

Synonyms: CAS: 101-21-3 ✦ BUD-NIP ✦ PREVENTOL ✦ TATERPEX ✦ CHLORPROPHAME

CHLOROPHYLL

Products and Uses: The green pigment in plants that is essential to photosynthesis. Recognized as a coloring and deodorant ingredient in personal care products. Used in casings, fats (rendered), margarine, shortening, soaps, tooth-pastes, cleaning products, chewing gum, waxes, liquors, cosmetics, perfumes, and dental hygiene products.

Precautions: Safe, natural, green coloring matter present in all plants except fungi and bacteria.

Synonyms: CAS: 1406-65-1 ✦ BIOPHYLL ✦ DAROTOL ✦ DEODOPHYLL ✦ GREEN CHLOROPHYL

CHLOROPICRIN

Products and Uses: An ingredient in fumigants, insecticides, rat exterminator, and tear gas. Products include fungicides, pesticides, and rat poisons.

Precautions: Very toxic by swallowing and breathing. A strong lung irritant and lachrimatory gas (causes eye irritation and tearing.)

Synonyms: CAS: 76-06-2 ✦ NITROCHLOROFORM ✦ TRICHLORONITROMETHANE ✦ VOMITING GAS

CHROMIUM PICOLINATE

Products and Uses: A patented form of chromium, a trace metal that is promoted for weight loss. It is reported to destroy fat while maintaining muscle and increasing strength. Chromium does bind insulin to cell membranes and may play a role in how the body uses carbohydrates. Promoters state that people do not get enough chromium. Actually, chromium deficiency is rare.

Precautions: The FDA has concerns that several times the daily limit of chromium (200 mcg) could be harmful. Some adverse effects include irregular heart beats.

Synonyms: NONE AVAILABLE.

CIGARETTE ADDITIVES

Products and Uses: There are 4000 ingredients in cigarettes. Additives include yeast, wine, caffeine, beeswax, and chocolate. Other ingredients, or by-products, include ammonia, angelica extract (which is a carcinogen in animals), benzene, benzopyrene, butone, cadmium, cyanide, DDT, ethyl furoate, lead formaldehyde, methoprene (insecticide), napthalene, and tars, and so on.

Precautions: Most of the previously mentioned are toxic or carcinogenic (causes cancer).

Synonyms: NONE FOUND.

CINNAMALDEHYDE

Products and Uses: Utilized in bakery products, beverages (nonalcoholic) colas, fruit, liquors, rum, spices, chewing gum, condiments, confections, ice cream products, meat, mouthwash, and toothpaste. Also in flavors and perfumery.

Precautions: Moderately toxic by swallowing excessive amounts. A possible allergen. A severe human skin irritant. GRAS (generally recognized as safe) when used in moderate amounts.

Synonyms: CAS: 104-55-2 ✦ BENZYLIDENEACETALDEHYDE ✦ CASSIA ALDEHYDE ✦ CINNAMAL ✦ CINNAMYL ALDEHYDE ✦ CINNIMIC ALDEHYDE ✦ PHENYLACROLEIN

CINNAMIC ACID

Products and Uses: Derived from cinnamon oil. A flavoring and odorant used in beverages, ice cream, candies, bakery goods, chewing gum, and suntan products. This is an ingredient that affects the taste and/or smell of final product.

Precautions: In excessive amounts it is moderately toxic by swallowing. A possible allergen. A skin irritant. Harmless when used in moderate amounts.

Synonyms: CAS: 621-82-9 ✦ PHENYLACRYLIC ACID ✦ tert-β-PHENYLACRYLIC ACID ✦ 3-PHENYLACRYLIC ACID ✦ 3-PHENYLPROPENOIC ACID ✦ 3-PHENYL-2-PROPENOIC ACID

CINNAMON LEAF OIL

Products and Uses: Found in bakery products, beverages (nonalcoholic), chewing gum, condiments, ice cream, meat, and pickles. As a flavoring for perfume and toothpaste.

Precautions: Possible allergen. FDA states GRAS (generally recognized as safe) when used in moderate amounts.

Synonyms: CINNAMON LEAF OIL, CEYLON ✦ CINNAMON LEAF OIL, SEYCHELLES

CINNAMYL ALCOHOL

Products and Uses: In soaps and cosmetics it is used as a scenting agent in perfumery, particularly for lilac and other floral scents.

Precautions: Moderately toxic by swallowing. A skin irritant.

Synonyms: CAS: 104-54-1 ✦ CINNAMIC ALDEHYDE ✦ CINNAMALDEHYDE ✦ 3-PHENYL PROPENAL ✦ 3-PHENYLALLYL ALCOHOL ✦ STYRONE ✦ STYRYL CARBINOL

CITRIC ACID

Products and Uses: An acid present in citrus fruits and berries. In the food industry there are multiple uses including beef (cured), chili con carne, cured meat food products, fats (poultry), fruits (frozen), lard, meat (dried), pork (cured), pork (fresh), potato sticks, potatoes (instant), poultry, sausage (dry), sausage (fresh pork), shortening, wheat chips, and wine. Used as a preservative, flavoring agent, sequestrant (affects the appearance, flavor or texture of final product), neutralizes lye in vegetable peelings, in curing of meats, prevents darkening of cut fruits and vegetables. A tart flavoring. A buffer that controls the acidity of jams, ice desserts, and candies.

Precautions: Mildly toxic by swallowing excessive amounts. A severe eye and moderate skin irritant, some allergenic effects. GRAS (generally recognized as safe) when used within normal limitations. USDA states limitations of use.

Synonyms: CAS: 77-92-9 ✦ ACILETTEN ✦ CITRETTEN ✦ CITRO

CITRONELLAL

Products and Uses: A flavoring agent used in bakery products, beverages (nonalcoholic), chewing gum, confections, gelatin desserts, ice cream, meat, and puddings.

Precautions: Use at a moderate level to accomplish the desired results.

Synonyms: CAS: 106-23-0 ✦ 3,7-DIMETHYL-6-OCTENAL

CITRONELLA OIL

Products and Uses: Found in insect repellent, soap, candles, perfumery, and disinfectants. It is derived from a fragrant grass grown in India.

Precautions: May not be used on food crops.

Synonyms: CAS: 8000-29-1

CLARY OIL

Products and Uses: Derived from a European herb. A food and beverage spice found in vermouth wine, root beer, cream soda, butter, fruits, licorice, black cherry and grape flavorings, ice desserts, candies, and bakery goods. Also used as a perfume fixative.

Precautions: GRAS (generally recognized as safe) when used in moderate amounts.

Synonyms: CLARY SAGE OIL ✦ OIL OF MUSCATEL

CLOVE LEAF OIL MADAGASCAR

Products and Uses: Used as a food flavoring agent, a dental analgesic (kills pain), and a dental germicide. Found in bakery products, beverages (alcoholic), beverages (nonalcoholic), root beer, cinnamon flavorings, cakes, chewing gum, condiments, confections, cookies, fruit punches, fruits (spiced), gelatin desserts, ice cream, marinades, meat, meat sauces, pickles, puddings, relishes, and sauces.

Precautions: In large amounts it is moderately toxic by swallowing and skin contact. A severe skin irritant and possible allergen. FDA states GRAS (generally recognized as safe) when used in moderate amounts.

Synonyms: CAS: 8015-97-2 ✦ CLOVE LEAF OIL ✦ OILS, CLOVE LEAF

COAL TAR

Products and Uses: Tar obtained from coal. Produced in making coke for steel. Coal tar originally was the major source of organic chemicals, which are currently obtained from natural gas and petroleum. Used for roads, waterproofing, paints, pipe coatings, surfacings, insulation, pesticides, sealants, adhesives, hair coloring products, and makeup.

Precautions: A human carcinogen. Toxic by breathing. Frequently causes skin irritation and allergic reactions.

Synonyms: CAS: 8007-45-2 ✦ COAL TAR PITCH VOLATILES ✦ LAVATAR ✦ POLYTAR BATH ✦ SYNTAR

COCAINE

Products and Uses: Used in medicine as a local narcotic anesthetic. Possession is illegal in U.S.

Precautions: A poison; central nervous system effects by swallowing. Other routes of consumption result in hallucinations, distorted perceptions, possible cardiac failure, and convulsions. An eye irritant. An abused controlled substance. Use leads to addiction and habit.

Synonyms: CAS: 50-36-2 ✦ METHYL BENZOYLECGONINE ✦ COKE ✦ CORINE ✦ ERITROXILINA ✦ GOLD DUST ✦ NEUROCAINE

COCOA BUTTER

Products and Uses: Derived from seeds of cocoa plant. In emollient creams, eyelash mascara removers, lipstick, nail cuticle products, blushes, soaps, chocolate, creams, ointments, and medicinal suppositories. As skin softener and lubricant.

Precautions: A possible allergen.

Synonyms: THEOBROMA OIL

COCONUT OIL

Products and Uses: Found in margarine, shortenings, deep-frying oil, synthetic cocoa butter, soaps, cosmetics, synthetic detergents, baby soaps, hair products, and cream bases.

Precautions: High fat content makes it undesirable for those watching cholesterol levels. Possible allergen. Generally harmless when used for intended purpose.

Synonyms: CAS: 8001-31-8 ✦ COPRA OIL ✦ COCONUT PALM OIL ✦ COCONUT BUTTER

COENZYME Q10

Products and Uses: Promoted as a health aid that inhibits the aging process along with other desirable results. This chemical is produced in virtually every cell of the body. This substance helps convert food into energy. It is also an antioxidant.

Precautions: The major disagreement is over whether it is desirable to consume it orally, or whether it is effective only when produced by the body itself.

Synonyms: NONE KNOWN.

COGNAC OIL

Products and Uses: Derived from the dregs of pressed grape skins used in winemaking. A flavoring agent in liquors, beverages, ice desserts, candies, bakery goods, and condiments.

Precautions: Harmless when used for intended purpose.

Synonyms: COGNAC OIL, WHITE ✦ COGNAC OIL, GREEN ✦ ETHYL OENANTHATE ✦ WINE YEAST OIL

COMFREY

Products and Uses: A root herb sold as a tea, tablet, capsule, tincture, poultice, or lotion. Promoted as having curative properties.

Precautions: Comfrey taken orally has been linked to liver damage (cirrhosis). Unborn children have suffered liver damage as a result of mother's use. Animal studies show lung, kidney, and gastrointestinal problems resulted from consumption. Australia, Canada, Germany, and Great Britain restrict sales of comfrey.

Synonyms: NONE KNOWN.

CONJUGATED LINOLEIC ACID

Products and Uses: A fatty acid that occurs naturally in meat and dairy products.

Precautions: No negative side effects known. In some lab tests, high doses of CLA prevented some cancers and heart disease, boosted immunities, and reduced body fats.

Synonyms: CLA

COPPER ACETORARSENITE

Products and Uses: Found in wood preservatives, larvicides, and antifouling paints

Precautions: Toxic by swallowing.

Synonyms: CUPRIC ACETOARSENITE ✦ PARIS GREEN ✦ KING'S GREEN ✦ SCHWEINFURT GREEN ✦ IMPERIAL GREEN

CORIANDER OIL

Products and Uses: A flavoring agent derived from the seed of the plant. Used in gin, curry powder, meat, sausage, toothpastes, raspberry flavorings, spice, ginger ale, candies, root beer, and condiments.

Precautions: A possible allergen. FDA states GRAS (generally recognized as safe) when used at moderate levels. May be irritating to the mouth when eaten, as it is a skin irritant.

Synonyms: CAS: 8008-52-4 ✦ OIL OF CORIANDER

CORK

Products and Uses: The outer bark of the cork oak tree. A renewable natural resource as the bark can be harvested every eight to ten years without harming the tree. Used as insulation for heat as well as noise; resistant to rot and mold. Also used for wall tiles. A component in linoleum floor tiles.

Precautions: Recognized as safe when used appropriately.

Synonyms: NONE KNOWN.

COSMECEUTICAL

Products and Uses: Refers to cosmetic products that have druglike benefits.

Precautions: The FDA does not recognize this term. Cosmetics are not approved by the FDA prior to sale. Drugs are products that cure, treat, mitigate, or prevent disease, or that affect the structure or function of the human body, according to the FDA.

Synonyms: NONE KNOWN.

COTTONSEED OIL

Products and Uses: Derived from the seed of the plant. Used in leather dressing, soap stock, base for cosmetics, nail polish remover, waterproofing products, dietary supplement, candy (hard and soft chocolate), salad dressings, margarine, mayonnaise, deep-frying oil, lards, and cooking oils. Acts as a lubricant in cosmetic and cooking products.

Precautions: Use at a moderate level not in excess of the amount required. Caused tumors and abnormal fetus development in animals. Frequent cause of allergic reactions.

Synonyms: CAS: 8001-29-4 ✦ DEODORIZED WINTERIZED COTTONSEED OIL

COUMARONE-INDENE RESIN

Products and Uses: As a protective coating for grapefruit, lemons, limes, oranges, tangelos, tangerines, in adhesives, and printing inks. Extends shelf life of fruits.

Precautions: FDA limits use on weight basis of fruit on which it is used. Fruits should be washed well before using.

Synonyms: NONE FOUND.

CREOSOTE, COAL-TAR

Products and Uses: Preservative for exterior items such as railroad ties, in wood treatment, foundation coatings, waterproofing, telephone poles, and waterside dock pilings. Used in fungicides, biocides, skin disinfectant, and veterinary skin treatments.

Precautions: A confirmed carcinogen (causes cancer). Toxic by breathing pungent fumes. Moderately toxic by ingestion. A skin and eye irritant.

Synonyms: CAS: 8001-58-9 ✦ CREOSOTE OIL ✦ LIQUID PITCH OIL ✦ TAR OIL ✦ BRICK OIL ✦ CRESYLIC CREOSOTE

CROTONALDEHYDE

Products and Uses: Found in insecticides, tear gases, fuel-gas components, and leather tanning. A solvent for varnishes and resins.

Precautions: Irritating to eyes and skin. A lachrymator (causes eyes to water). Flammable, a fire risk.

Synonyms: CAS: 123-73-9 ✦ 2-BUTENAL ✦ CROTONIC ✦ ALDEHYDE

CUPROUS IODIDE

Products and Uses: Dietary supplement added to table salt to prevent goiters. Source of iodine in table salt.

Precautions: GRAS (generally recognized as safe) with a limitation of 0.01% in table salt.

Synonyms: CAS: 7681-65-4 ✦ COPPER IODIDE ✦ COPPER(I) IODIDE

CYCLAMATE

Products and Uses: Sweetener (nonnutritive). Approved for use in Europe and in 40 countries around the world. Currently under study in the U.S.

Precautions: This was a popular sweetener in the 1960s until it was removed from the market in 1969. Suspected human carcinogen (may cause cancer), produced bladder tumors. Mildly toxic by swallowing excessive amounts.

Synonyms: CAS: 100-88-9 ✦ CYCLAMIC ACID ✦ CYCLOHEXANESULPHAMIC ACID ✦ CYCLOHEXYLAMIDOSULPHURIC ACID ✦ CYCLOHEXYLAMINESULPHONIC ACID ✦ CYCLOHEXYLSULFAMIC ACID ✦ CYCLOHEXYLSULPHAMIC ACID ✦ HEXAMIC ACID ✦ SUCARYL ✦ SUCARYL ACID

CYCLOHEXANE

Products and Uses: Used as a solvent in paint and varnish remover. Also used in solid fuels and fungicides.

Precautions: Moderately toxic by breathing and skin contact.

Synonyms: CAS: 110-87-7 ✦ HEXAMETHYLENE ✦ HEXANAPHTHENE ✦ HEXALHYDROBENZENE

p-CYMENE

Products and Uses: Flavoring for foods and odorant fragrance for spices and citrus flavorings. It is found in nearly 100 volatile oils including lemongrass, sage, thyme, coriander, star anise, and cinnamon. Products include beverages, cream desserts, candy, and bakery goods

Precautions: Mildly toxic by swallowing large amounts. Humans sustain central nervous system effects at low doses. A skin irritant. FDA approves use at modest levels to accomplish the intended effect.

Synonyms: CAS: 99-87-6 ✦ CAMPHOGEN ✦ CYMENE ✦ CYMOL ✦ DOLCYMENE ✦ PARACYMENE ✦ PARACYMO

L-CYSTINE

Products and Uses: An ingredient in baked goods (yeast leavened) and baking mixes. Considered as a dietary supplement, dough strengthener, and nutrient.

Precautions: GRAS (generally recognized as safe) when used within FDA limitations.

Synonyms: CAS: 56-89-3 ✦ CYSTEINE DISULFIDE ✦ CYSTIN ✦ (-)-CYSTINE ✦ CYSTINE ACID ✦ DICYSTEINE ✦ β,β'-DITHIODIALANINE ✦ GELUCYSTINE ✦ OXIDIZED l-CYSTEINE

2,4-D

Products and Uses: A fungicide, herbicide, weed killer, and defoliant that controls fruit dropping. It is used on barley (milled fractions, except flour), oats (milled fractions, except flour), potable water, rye (milled fractions), sugarcane bagasse, sugarcane molasses, and wheat (milled fractions, except flour).

Precautions: A suspected human carcinogen (may cause cancer). Poison by swallowing. Moderately toxic by skin contact. Effects on the human body by swallowing are somnolence (sleepiness), convulsions, coma, and nausea or vomiting. Can cause liver and kidney injury. A skin and severe eye irritant.

Synonyms: CAS: 94-75-7 ✦ AGROTECT ✦ AQUA-KLEEN ✦ CHLOROXONE ✦ CROP RIDER ✦ DED-WEED ✦ ESTERON BRUSH KILLER ✦ FARMCO ✦ HERBIDAL ✦ LAWN-KEEP ✦ MIRACLE ✦ PLANOTOX ✦ PLANTGARD ✦ SALVO ✦ SUPER D WEEDONE ✦ WEED-B-GON ✦ WEEDEZ WONDER BAR ✦ WEEDONE LV4 ✦ WEED TOX ✦ WEEDTROL

DALAPON

Products and Uses: A herbicide that kills plants and foliage.

Precautions: Strong irritant to eyes and skin.

Synonyms: CAS: 75-00-0 ✦ 2,2-DICHLOROPROPIONIC ACID

DAMINOZIDE

Products and Uses: A plant growth regulator applied to apples and other orchard crops. Used on cherries, nectarines, pears, grapes, peanuts, plums, and peaches. It increases the shelf life, coloring and firmness for apple, cherry, prune, pear, grape, nectarine, and peanut crops. It increases storage life and preserves firmness and coloration.

Precautions: The chemical penetrates the whole fruit and cannot be washed off or removed by peeling. A carcinogen (causes cancer) in lab animals.

Synonyms: ALAR

DDH

Products and Uses: A common disinfectant, antiseptic, whitening agent, and purifying agent found in household laundry bleach, water treatments, swimming pool treatments, and mild chlorinating agents.

Precautions: Mildly toxic by swallowing and breathing. A severe skin irritant. A mutagen (changes inherited characteristics). Avoid contact because of effects of active chlorine on skin. Sometimes these chemicals are central nervous system depressants (slows heart rate and breathing rate).

Synonyms: CAS: 118-52-5 ✦ DICHLORANTIN ✦ DANTOIN ✦ DACTIN ✦ HYDAN ✦ HALANE ✦ DCA

1-DECANAL

Products and Uses: Found in over 50 sources including citrus oils, citronella, and lemongrass. A flavoring agent used as fruit and berry flavoring in beverages, ice desserts, candies, bakery products, gum, and gelatins.

Precautions: Moderately toxic by swallowing large amounts. Mildly toxic by skin contact when applied to the skin for extended time. Avoid overuse.

Synonyms: CAS: 112-31-2 ✦ ALDEHYDE C10 ✦ C-10 ALDEHYDE ✦ CAPRALDEHYDE ✦ 1-DECYL ALDEHYDE

DEDECYL ALCOHOL

Products and Uses: In baked goods, beverages, candy, ice cream, butter, coconut, and fruit. Also an odorant in detergents, lubricants, solvents, and perfumes. Considered a flavoring agent.

Precautions: Moderately toxic by skin contact. In gross amounts it is mildly toxic by swallowing and breathing. A severe skin and eye irritant. Possible carcinogen (may cause cancer) that produced tumors in animals. Use at a level not in excess of the amount required to accomplish the desired results.

Synonyms: CAS: 112-30-1 ✦ ALCOHOL C-10 ✦ ANTAK ✦ CAPRIC
ALCOHOL ✦ DECANAL DIMETHYL ACETAL ✦ DECANOL

DEET

Products and Uses: Insect repellents, solvent, and film former. Various uses, primarily as a pesticide or repellent.

Precautions: Toxic by swallowing. Irritant to eyes, nose and throat. Usually 30% concentration or less is effective and not toxic to humans. Avoid using products that contain DEET on cats. Cats are more vulnerable to toxic substances than humans or dogs because they lack the enzymes that break down toxic substances in the liver.

Synonyms: CAS: 134-62-3 ✦ N,N-DIETHYL-m-TOLUAMIDE ✦ DELPHENE ✦
FLYPEL ✦ OFF ✦ REPEL

DEHYDROACETIC ACID

Products and Uses: Utilized in fungicides, bactericides, medicated toothpastes, tooth powders, makeup, and shampoos. It destroys fungi and bacteria in dentifrices and cosmetics.

Precautions: Toxic by swallowing.

Synonyms: CAS: 520-45-6 ✦ DHA ✦ METHYLACETOPYRANONE ✦
DEHYDRACETIC ACID ✦ SODIUM DEHYDROACETATE

DEICER CHEMICALS

Products and Uses: Produced in the form of flakes, granules or pellets to melt ice and snow from walkways or streets. In order to work, they require moisture to dissolve and form brine solutions that lower the freezing point of water. Their brine solutions melt ice and snow and penetrate down to pavement, undercutting the ice from the pavement. Calcium chloride is the only deicer that gives off heat as it dissolves. This works faster than other types.

Precautions: Harmless when used according to package directions using appropriate caution. Methanol and alcohol are frequently active ingredients. Precautions must be observed.

Synonyms: CALCIUM CHLORIDE ✦ SODIUM CHLORIDE ✦ ROCK SALT ✦
POTASSIUM CHLORIDE ✦ UREA

DEMETON-S

Products and Uses: An ingredient in pesticides. A systemic insecticide that is absorbed by the plant, which then becomes toxic to sucking and chewing insects.

Precautions: Poison by swallowing.

Synonyms: CAS: 126-75-0 ✦ DIETHYL-S-(2-ETHIOETHYL)THIOPHOSPHATE ✦ O,O-DIETHYL-S-2-(ETHYLTHIO)ETHYL PHOSPHOROTHIOATE ✦ O,O-DIETHYL-S-(2-(ETHYLTHIO)ETHYL) PHOSPHOROTHIOLATE (USDA) ✦ ISODEMETON ✦ PO-SYSTOX ✦ THIOLDEMETON ✦ SYSTOX

DENATURED ALCOHOL

Products and Uses: At least 50 formulations are authorized for making denatured alcohol. Used for solvents, antifreeze, and brake fluids.

Precautions: Flammable. A dangerous fire risk.

Synonyms: DENATURED SPIRITS

DEXTRINS

Products and Uses: Found in baked goods, beverages (dry mix), confectionery products, egg roll, food-contact surfaces, gravies, pie fillings, poultry, puddings, and soups. Its purpose is as a binder, stabilizer, extender in adhesives and printing inks, a surface-finishing agent for paper and textiles, and also a thickening agent.

Precautions: GRAS (generally recognized as safe) when used at moderate levels for intended purposes. Possible allergen.

Synonyms: CAS: 9004-53-9 ✦ ARTIFICIAL GUM ✦ DEXTRANS ✦ STARCH GUM ✦ TAPIOCA ✦ VEGETABLE GUM

DHEA

Products and Uses: Currently a very popular supplement being sold in drugstores, and health food stores. It is called the "mother hormone," and is touted as a cancer preventative, fountain of youth, heart disease and AIDS cure. These are all unproven.

Precautions: The long-term effects on the body are unknown. The many claims are unproven. The over-the-counter product is sometimes an extract of wild yams, which is supposed to convert to DHEA in the body.

Synonyms: DEHYDROEPIANDROSTERONE

DIALIFOR

Products and Uses: Insecticide for apple pomace (dried), citrus pulp (dried), grape pomace (dried), raisin waste, and raisins. Used as acaricide (mite killer), pesticide against moths and red spider mite of fruit.

Precautions: Poison by swallowing and skin contact. An animal teratogen (abnormal fetus development). Has other animal reproductive effects (infertility or sterility or birth defects). FDA states specific limitations for use on food products.

Synonyms: CAS: 10311-84-9 ✦ S-(2-CHLORO-1-PHTHALIMIDOETHYL)-O,O-DIETHYL PHOSPHORODITHIOATE ✦ O,O-DIETHYL-S-(2-CHLORO-1-PHTHALIMIDOETHYL)PHOSPHORODITHIOATE ✦ TORAK

DIAMYL PHENOL

Products and Uses: Found in oils, rust preventatives, and detergent products for the purpose of lubrication and cleaning.

Precautions: Irritant to skin. Could cause allergic reaction.

Synonyms: 1-HYDROXY-2,4-DIAMYLBENZENE

DIATOMACEOUS EARTH

Products and Uses: A filler in paints, tooth polish, cosmetics, and facial powder. Also used as an abrasive, absorber, filter aid, and anticaking agent.

Precautions: A nuisance dust that may cause fibrosis of the lungs. A possible carcinogen (may cause cancer). Composed of the skeletons of small aquatic plants.

Synonyms: CAS: 61790-53-2 ✦ D.E. ✦ DIATOMACEOUS SILICA ✦ DIATOMITE ✦ INFUSORIAL EARTH ✦ KIESELGUHR

DIAZINON

Products and Uses: Insecticide. As a pesticide that is particularly effective against fire ants. This use permitted by EPA.

Precautions: Poison by swallowing and skin contact. Mildly toxic by breathing. Effects on the body by swallowing are changes in motor activity, muscle weakness, and sweating. Avoid cat flea collars with this ingredient, as it is harmful to animals. Animal teratogenic (abnormal fetus development) and has reproductive effects (infertility or sterility or birth defects). A skin and severe eye irritant. A human mutagen (changes inherited characteristics).

Synonyms: CAS: 333-41-5 ✦ ALFA-TOX ✦ DIANON ✦ DIAZIDE ✦ DIAZINONE ✦ DIAZITOL ✦ DIAZOL ✦ DIZINON ✦ GARDENTOX ✦ SPECTRACIDE

DIBENZOTHIOPHENE

Products and Uses: Ingredient frequently found in antidandruff shampoos.

Precautions: Not for internal use as it is moderately toxic.

Synonyms: CAS: 132-65-0 ✦ 9-THIAFLUORENE ✦ DIPHENYLENE SULFIDE

DIBENZYL DISULFIDE

Products and Uses: For lubrication in petroleum oils and greases.

Precautions: Moderately toxic by swallowing. A skin and eye irritant.

Synonyms: CAS: 150-60-7 ✦ BENZYL DISULFIDE

DIBENZYL ETHER

Products and Uses: A synthetic spice and fruit flavoring found in beverages, ice desserts, candies, bakery goods, and gums.

Precautions: Moderately toxic by swallowing gross amounts. Vapors are probably narcotic in high concentration. A skin and eye irritant.

Synonyms: CAS: 103-50-4 ✦ BENZYL ETHER

DIBROMODIFLUOROMETHANE

Products and Uses: A fire extinguishing agent and a direct-contact freezing agent for foods.

Precautions: Mildly toxic by breathing.

Synonyms: CAS: 75-61-6 ✦ DIFLUORODIBROMOMETHANE ✦ FREON 12-B2 ✦ HALON 1202

DIBROMOFLUORESCEIN

Products and Uses: An ingredient in lipstick, blushes, makeup, as a red-orange coloring or dye.

Precautions: A possible allergen and stomach irritant upon swallowing. Possible skin irritant.

Synonyms: NONE FOUND.

DIBROMOPROPANOL

Products and Uses: Used as a coating for flame retardant products.

Precautions: A carcinogen.

Synonyms: CAS: 96-13-9 ✦ 2,3-DIBROMO-1-PROPANOL

DIBUTYL BUTYL PHOSPHONATE

Products and Uses: An ingredient in fabric softener used as a textile conditioner and antistatic agent.

Precautions: A possible allergen.

Synonyms: NONE FOUND.

DIBUTYL PHTHALATE

Products and Uses: Found in cosmetics, inks, perfumes, and glues. Also used as a plasticizer in nail polish, and in printing inks, paper coatings, and insect repellent for textiles.

Precautions: Mildly toxic by swallowing. Effects on the human body by swallowing are hallucinations, distorted perceptions, nausea or vomiting, kidney, ureter, or bladder changes. A possible mutagen (changes inherited characteristics).

Synonyms: CAS: 84-74-2 ✦ DBP ✦ CELLUFLEX ✦ PALATINOL ✦ 1,2-BENZENE-DICARBOXYLATE

DIBUTYL SEBACATE

Products and Uses: A plasticizer in rubber softeners that is also used for fruit-flavored cosmetics and perfumes.

Precautions: Mildly toxic by swallowing. Animal reproductive effects (infertility or sterility or birth defects).

Synonyms: CAS: 109-43-3 ✦ DECANEDIOIC ACID, DIBUTYL ESTER ✦ DI-n-BUTYL SEBACATE ✦ KODAFLEX DBS ✦ MONOPLEX DBS ✦ POLYCIZER DBS ✦ SEBACIC ACID, DIBUTYL ESTER ✦ STAFLEX DBS

DICHLORAMINE-T

Products and Uses: In antiseptic medication. A germicide and antibacterial.

Precautions: A possible allergen.

Synonyms: CHLORAMINE-T ✦ SODIUM p-TOLUENESULFOCHLOR ✦ p-TOLUENESULFONDICHLORAMIDE

DICHLOROBENZALKONIUM CHLORIDE

Products and Uses: Additive in agricultural products as an algicide (kills algae), antiseptic, and sterilizing agent.

Precautions: A deadly poison by swallowing. A severe eye irritant. Can cause liver and kidney damage. Could cause allergic reaction.

Synonyms: CAS: 8023-53-8 ✦ TETROSAN

o-DICHLOROBENZENE

Products and Uses: Found in fumigants, insecticides, vehicle polishes, air deodorants (fresheners), shoe dyes, shoe polishes, tar removers, and grease removers as a pesticide, polish, odorant, and solvent.

Precautions: Poison by swallowing. Moderately toxic by breathing. An animal teratogen (abnormal fetus development), which also caused reproductive effects (infertility or sterility or birth defects). An eye, skin, nose, and throat irritant. A possible carcinogen (may cause cancer). A mutagen (changes inherited characteristics).

Synonyms: CAS: 95-50-1 ✦ 1,2-DICHLOROBENZENE ✦ CHLOROBEN ✦ CHLORODEN ✦ DCB ✦ TERMITKIL

p-DICHLOROBENZENE

Products and Uses: As a fumigant, moth repellent, germicide, space odorant, and soil fumigant. As a pesticide, bactericide, and air deodorant.

Precautions: A definite carcinogen (causes cancer). An animal teratogen (abnormal fetus development). Moderately toxic to humans by swallowing. Various effects on the body by swallowing include liver changes, respiratory effects, and constipation. An eye irritant.

Synonyms: CAS: 106-46-7 ✦ 1,4-DICHLOROBENZENE ✦ PARACIDE ✦ PARADICHLORBENZENE ✦ PARAMOTH ✦ PARAZENE ✦ PDB ✦ PDBC

DICHLORODIFLUOROMETHANE

Products and Uses: Its purpose is as a refrigerant in air conditioners, for freezing of foods by direct contact, and the chilling of cocktail glasses.

Precautions: Effects on the body by breathing include eye irritation, lung irritation, and liver changes. A narcotic in high concentrations.

Synonyms: CAS: 75-71-8 ✦ FLUOROCARBON 12 ✦ ARCTON ✦ FREON F 12 ✦ PROPELLANT 12 ✦ HALON

DICHLOROETHER

Products and Uses: Found in paints, varnishes, lacquers, finish removers, spot remover, and dry cleaning fluid. Used as a solvent and in penetrating compounds.

Precautions: A poison by swallowing, skin contact, and breathing. A skin, eye, nose, and throat irritant. Questionable carcinogen (may cause cancer).

Synonyms: CAS: 111-44-4 ✦ DICHLOROETHYL ETHER ✦ CHLOREX ✦ DCEE

α-DICHLOROHYDRIN

Products and Uses: General solvent for paints, varnishes, and lacquers as a dissolver and binder in paint products.

Precautions: Toxic by breathing and swallowing.

Synonyms: DICHLOROISPROPYL ALCOHOL ✦ 1,3-DICHLORO-2-PROPANOL

DICHLOROISOCYANURIC ACID

Products and Uses: Utilized in household dry bleaches, dishwashing compounds, scouring powders, and detergent sanitizers. It is also a replacement for calcium hypochlorite.

Precautions: May ignite organic materials on contact. Irritant to eyes.

Synonyms: DICHLORO-S-TRIAZINE-2,4,6-TRIONE

DICHLOROISOPROPYL ETHER

Products and Uses: Ingredient in spot removers, dry-cleaning solutions, paint, and varnish remover. Also useful as a solvent for oils, greases, fats, and waxes.

Precautions: Moderately toxic by swallowing. Moderately toxic by skin contact and breathing. An eye irritant. Questionable carcinogen (may cause cancer). A corrosive material.

Synonyms: CAS: 108-60-1 ✦ BIS(2-CHLORO-1-METHLETHYL)ETHER ✦ DICHLORODIISOPROPYL ETHER

DICHLOROPENTANE

Products and Uses: A paint, varnish remover, and wax remover. It is a soil fumigant and a solvent for oil, grease, and tar.

Precautions: Flammable, dangerous fire risk.

Synonyms: CAS: 30586-10-8 ✦ CHLORINATED HYDROCARBONS, ALIPHATIC

DICHLOROPHENE

Products and Uses: Found in cosmetics, deodorants, shampoos, and toothpastes as a fungicide and germ killer.

Precautions: Moderately toxic by swallowing. A skin and severe eye irritant. Frequently causes allergic reaction. Can cause cramps and diarrhea.

Synonyms: CAS: 97-23-4 ✦ ANTIPHEN ✦ PANACIDE ✦ TENIATHANE ✦ WESPURIAL

DICHLORVOS

Products and Uses: In flea and tick collars, sprays, and roach and ant killers. An insecticide for animal feed, cereals, cookies (packaged), crackers (packaged), figs (dried), flour, pork, and sugar. Excellent for controlling spiders and flying insects.

Precautions: Dichlorvos works by affecting the central nervous system and is extremely hazardous. It is a suspected carcinogen (cancer causing chemical). The

U.K. Department of the Environment is considering banning this and all products containing same. Poison by swallowing gross amounts. Also poison by breathing and skin contact. FDA states limits on amounts used on human foods.

Synonyms: CAS: 62-73-7 ✦ CANOGARD ✦ CHLORVINPHOS ✦ CYANOPHOS ✦ DDVF ✦ DDVP ✦ DERIBAN ✦ DICHLOROPHOS ✦ FLY-DIE ✦ NO-PEST ✦ NO-PEST STRIP ✦ TASK TABS

DICYCLOHEXYLAMINE

Products and Uses: An ingredient in paints, varnishes, inks, and detergents.

Precautions: Toxic by swallowing, strong irritant to skin, nose, and throat.

Synonyms: CAS: 101-83-7 ✦ CDHA ✦ DODECAHYDRODOPHENYLAMINE

DIETHANOLAMINE

Products and Uses: Found in liquid detergents, shampoos, cleaners, and polishes. Uses include as dispersing agents, emollients (softener), and humectants (collects moisture).

Precautions: A poison by swallowing, skin contact, and breathing. A skin, eye, nose, and throat irritant. A possible carcinogen (may cause cancer).

Synonyms: CAS: 111-42-2 ✦ DEA ✦ DIOLAMINE ✦ DIETHYLOLAMINE

DIETHYLAMINOETHANOL

Products and Uses: A colorless liquid that has various uses including in fabric softeners and antirust compositions.

Precautions: Toxic by swallowing and skin absorption.

Synonyms: CAS: 100-37-8 ✦ DIETHYLETHANOLAMINE ✦ 2-HYDROXYTRIETHYLAMINE

DIETHYL DICARBONATE

Products and Uses: This chemical inhibits fermentation. Used as a fungicide.

Precautions: Poison by swallowing. Concentrated DEPC is irritating to eyes, nose, throat, and skin. Prohibited from direct addition or use in human food. Legal for use in wine in other countries.

Synonyms: CAS: 1609-47-8 ✦ BAYCOVIN ✦ DEPC ✦ DICARBONIC ACID DIETHYL ESTER ✦ DIETHYL ESTER of PYROCARBONIC ACID ✦ DIETHYL OXYDIFORMATE ✦ DIETHYL PYROCARBONIC ACID

DIETHYLENE ETHER

Products and Uses: Various applications include lacquer, paint, varnish, solvents and removers, cleaning preparations, detergent preparations, cements, cosmetics, and deodorants.

Precautions: Toxic by breathing. Absorbed by skin. A carcinogen (causes cancer).

Synonyms: CAS: 123-91-1 ✦ 1,4-DIOXANE ✦ DIETHYLENE OXIDE ✦ DIOXYETHYLENE ETHER ✦ 1,4-DIETHYLENE DIOXIDE

DIETHYLENE GLYCOL

Products and Uses: A conditioner, humectant (moisturizer), and cleaner used in fabric softeners, tobacco moisturizer, synthetic sponges, paper products, corks, book-binding adhesives, cosmetics, rug cleaners, upholstery cleaners, floor polish, furniture polish, as disinfectant, and antifreeze.

Precautions: Moderately toxic to humans by swallowing. A skin and eye irritant.

Synonyms: CAS: 111-46-6 ✦ DIHYDROXYDIETHYL ETHER ✦ DIGLYCOL ✦ DEG ✦ GLYCOL ETHYL ETHER

DIETHYL PHTHALATE

Products and Uses: A plasticizer (for flexibility) in wrapping and packaging materials. Substance is also found in mosquito repellent and other insecticidal sprays.

Precautions: Moderately toxic by swallowing. Effects by breathing are lachrymation, (watering eyes), respiratory obstruction, and other unspecified respiratory system effects. An eye irritant. Narcotic in high concentrations.

Synonyms: CAS: 84-66-2 ✦ ANOZOL ✦ DIETHYL-o-PHTHALATE ✦ ETHYL PHTHALATE ✦ NEANTINE ✦ PHTHALOL ✦ SOLVANOL

DIFLUORPHOSPHORIC ACID

Products and Uses: Considered a protective coating for metal. Found in polishes for vehicle surfaces and metal protectors.

Precautions: Could be corrosive to skin or eyes. An irritant.

Synonyms: PHOSPHORODIFLUORIDIC ACID

DIGITOXIN

Products and Uses: Prescribed as a medication for cardiac treatment.

Precautions: Derived from the foxglove plant. Can be toxic by swallowing excessive amounts and overdose can be fatal.

Synonyms: CAS: 71-63-6

DIGLYCOL LAURATE

Products and Uses: Commonly used in hand lotions, hair dressings, and dry-cleaning soap. It is an emulsifier (to aid in suspension of oily liquids) and prevents separation.

Precautions: Can cause allergic reactions.

Synonyms: DIETHYLENE GLYCOL MONOLAURATE

DIGLYCOL MONOSTEARATE

Products and Uses: Found in skin moisturizers, lotions, creams, cosmetics, and makeup. Used as an emulsifier (to aid in suspension of oily liquids) and as a thickener.

Precautions: Can cause allergic reactions.

Synonyms: CAS: 106-11-6 ✦ DIETHYLENE GLYCOL MONOSTEARATE

DIGLYCOL STEARATE

Products and Uses: Utilized as an emulsifier (to aid in suspension of oils), as a lubricant, and as a thickener or filler, in powder, polishes, and cleaners.

Precautions: Can cause allergic reactions.

Synonyms: DIETHYLENE GLYCOL DISTEARATE

DIHYDROXYACETONE

Products and Uses: Found in cosmetic products used to produce artificial suntan appearance on skin.

Precautions: Possible skin irritant and could cause allergic reactions.

Synonyms: DHA ✦ DIHYDROXYPROPANONE

2,5-DIHYDROXYBENZOQUINONE

Products and Uses: Utilized in skin products as a tanning agent.

Precautions: Derived from the chemical hydroquinone. An irritant to skin and eyes.

Synonyms: NONE FOUND.

5,7-DIHYDROXY-4-METHYLCOUMARIN

Products and Uses: Commonly used in suntan oil and lotion products as a sunscreen. Found in wall paints as a pigment and whitening agent. A UV (ultraviolet) light absorber.

Precautions: Possible cause of allergic reactions.

Synonyms: NONE FOUND.

DIHYDROXYSTEARIC ACID

Products and Uses: Utilized in cosmetics and lotions as a thickener and stabilizer in makeup products.

Precautions: Can cause allergic reactions.

Synonyms: NONE FOUND.

DIISOBUTYL KETONE

Products and Uses: Found in lacquers, inks, and stains as a solvent in paint-type products.

Precautions: Moderately toxic by swallowing and breathing. Mildly toxic by skin contact. An eye and skin irritant. Narcotic in high concentration. Breathing can cause headache, nausea, and vomiting.

Synonyms: CAS: 108-83-8 ✦ ISOVALERONE ✦ VALERONE

DIISOPROPANOLAMINE

Products and Uses: Ingredient in polishes, leather preservatives, leather polishes, oils, and water paints. An emulsifier (aids in suspension of oils), stabilizer, and maintains mixes of chemicals in products.

Precautions: Can cause allergic reactions.

Synonyms: DIPA

DIISOPROPYL DIXANTHOGEN

Products and Uses: Component in fungicide, as weed killer and herbicide.

Precautions: Toxic by swallowing and breathing. A strong irritant.

Synonyms: NONE FOUND.

DIKETENE

Products and Uses: A food preservative used to maintain food quality.

Precautions: In large amounts it is moderately toxic by swallowing. It is toxic by skin contact. A skin and severe eye irritant.

Synonyms: CAS: 674-82-8 ✦ ACETYL KETENE ✦ KETENE DIMER

DILL SEED OIL, INDIAN TYPE

Products and Uses: As a flavoring agent in dips, meats, sauces, and spreads.

Precautions: GRAS (generally recognized as safe).

Synonyms: DILL OIL, INDIAN TYPE ✦ DILL SEED OIL, INDIAN

DILL WEED OIL

Products and Uses: Found in bakery goods, sauces, meat sausages, and pickles as a seasoning.

Precautions: In large amounts it could be mildly toxic by swallowing. A skin irritant.

Synonyms: CAS: 8006-75-5 ✦ DILL FRUIT OIL ✦ DILL HERB OIL ✦ DILL OIL ✦ DILL SEED OIL

DIMETHOXANE

Products and Uses: Various applications include as a preservative for cosmetics and inks and a gasoline additive. A chemical that stabilizes and preserves products.

Precautions: Possible carcinogen (may cause cancer). Moderately toxic by swallowing.

Synonyms: CAS: 828-00-2 ✦ ACETOMETHOXANE ✦ DIOXIN BACTERICIDE

N,N-DIMETHYL ACETAMIDE

Products and Uses: A solvent in paint remover.

Precautions: Toxic by breathing. It is absorbed by skin. A strong irritant.

Synonyms: CAS: 127-19-5 ✦ DMAC

DIMETHYLAMINOETHYL METHACRYLATE

Products and Uses: A fabric antistatic agent used as a fabric softener or coating agent for textiles.

Precautions: An irritant to skin, eyes, nose, and throat. A strong lachrymator (causes eyes to water).

Synonyms: NONE FOUND.

2,5-DIMETHYLBENZYL CHLORIDE

Products and Uses: Found in dyes, perfumes, and germicides.

Precautions: Irritant to eyes, nose, and throat. A lachrymator (causes eyes to water).

Synonyms: α-CHLORO-p-XYLENE

DIMETHYL CHLOROTHIOPHOSPHATE

Products and Uses: Utilized in flame retardant coatings, sprays, fungicides, and pesticides for camping equipment and tent fabric coatings, insecticides, and additives.

Precautions: Poison by breathing. Moderately toxic by swallowing and skin contact. Corrosive. Use may be restricted.

Synonyms: CAS: 2524-03-0 ✦ METHYL PCT ✦ DIMETHYL PHOSPHOROCHLORIDOTHIOATE

DIMETHYL DICARBONATE

Products and Uses: Found in wine as a fungicide and yeast inhibitor.

Precautions: FDA requires limitation of 200 ppm in wine.

Synonyms: CAS: 4525-33-1

DIMETHYLHYDANTOIN-FORMALDEHYDE POLYMER

Products and Uses: A fixative in adhesives and aerosol hair sprays.

Precautions: Avoid breathing.

Synonyms: NONE FOUND.

2,6-DIMETHYLMORPHOLINE

Products and Uses: Utilized in self-polishing floor polishes, germicides, and textile finishes.

Precautions: Moderately toxic by swallowing and skin contact. A skin irritant.

Synonyms: CAS: 141-91-3

DIMETHYLOCTANOL

Products and Uses: An ingredient in bakery products, beverages (nonalcoholic), chewing gum, confections, ice cream products, and pickles as a floral and fruit flavoring agent.

Precautions: Moderately toxic by skin contact. A skin irritant.

Synonyms: CAS: 106-21-8 ✦ DIHYDROCITRONELLOL ✦ GERANIOL TETRAHYDRIDE ✦ PELARGOL ✦ PERHYDROGERANIOL ✦ TETRAHYDROGERANIOL

DIMETHYLOLUREA

Products and Uses: A resin used in the formation of plywood and pressboard lamination. Its purpose is to increase the hardness and fire resistance in wood products.

Precautions: Avoid breathing of vapors as degassing occurs.

Synonyms: DMU ✦ 1,3-BISHYDROXYMETHLUREA

DIMETHYL PHTHALATE

Products and Uses: A coating agent to aid flexibility in lacquers, plastics, and rubber products.

Precautions: Moderately toxic by swallowing and by breathing. An irritant to eyes, nose, and throats. Not absorbed by the skin.

Synonyms: CAS: 131-11-3 ✦ AVOLIN ✦ DMP ✦ METHYL PHTHALATE ✦ PHTHALIC ACID METHYL ESTER

DIMETHYLPOLYSILOXANE

Products and Uses: Various applications include as a defoamer and release agent in chewing gum, gelatins, poultry, salt, sugar, wine, transformer liquid, and brake fluids.

Precautions: FDA approves use within limitations.

Synonyms: DIMETHYL SILICONE ✦ POLYDIMETHYLSILOXANE

DIMETHYL SULFIDE

Products and Uses: A gas odorant that gives natural gas an identifiable odor.

Precautions: Flammable, dangerous fire risk, moderate explosion risk.

Synonyms: CAS: 75-18-3 ✦ METHYL SULFIDE

DIMETHYL SULFOXIDE

Products and Uses: Found in industrial cleaners, pesticides, and paint-stripping solvents. It is believed by some that it is an effective anti-inflammatory for arthritic conditions. Not recommended by health professionals.

Precautions: Poison by swallowing. A skin and eye irritant. A human mutagen (affects inherited characteristics). Can cause anaphylactic reaction and corneal opacity. Readily penetrates skin and other tissues.

Synonyms: CAS: 67-68-5 ✦ DMSO ✦ METHYL SULFOXIDE

DIMYRISTYL ETHER

Products and Uses: An ingredient in water repellent sprays and coatings used as an antistatic fabric softener, coating, or conditioner.

Precautions: Could cause allergic reactions on skin contact.

Synonyms: DITETRADECYL ETHER

2,4-DINITROANISOLE

Products and Uses: Found in insecticides and pesticides for moth, carpet beetle, cockroach, and in body lice repellent.

Precautions: Toxic by swallowing.

Synonyms: 2,4-DINITRIOPHENYL METHYL

DINITROPHENOL

Products and Uses: Various uses include in dyes, lumber preservatives, and photographic developers.

Precautions: Wear gloves when handling lumber treated with preservative. Chemical can pass through and be absorbed by unbroken skin. Breathing of dust could be fatal.

Synonyms: CAS: 25550-58-7

DINOSEB

Products and Uses: A gardening or agricultural chemical. An herbicide for pre-emergence treatment. Purpose is to prevent the sprouting of weeds.

Precautions: A possible fire risk. A strong irritant, absorbed by the skin.

Synonyms: CAS: 88-85-7 ✦ 2-SEC-BUTYL-4,6-DINITROPHENOL ✦ 2,4-DINITRO-6-SEC-BUTYLPHENOL

DIOXIN

Products and Uses: Among the most toxic synthetic chemicals known. It is produced when certain organic compounds are burned. It is also a byproduct of the chlorine bleaching process of wood pulp to produce papers, diapers, and so on. Sweden now avoids the problem by producing 95% of their paper products unbleached.

Precautions: Effluents from manufacturing plants using chlorine bleach seriously contaminate rivers, lakes, and seas.

Synonyms: DIBENZO-P-DIOXINS

DIPENTENE

Products and Uses: An ingredient in paints, enamels, lacquers, varnishes, printing inks, perfumes, flavors, floor waxes, and furniture polishes. Used as a solvent or wax.

Precautions: A skin irritant.

Synonyms: CAS: 138-86-3 ✦ CINENE ✦ LIMONENE, INACTIVE ✦ CAJEPUTENE ✦ DIPANOL ✦ KAUTSCHIN

DIPHENYLAMINE CHLOROARSINE

Products and Uses: A poison gas for self-defense, used as a crowd control gas; also as a wood preservative.

Precautions: Toxic by breathing and swallowing. A strong irritant.

Synonyms: CAS: 578-94-9 ✦ ADAMSITE ✦ PHENARSAZINE CHLORIDE ✦ DM

DIPHENYLMETHANE

Products and Uses: Found in various dyes and perfumery.

Precautions: Toxic by swallowing.

Synonyms: BENZYLBENZENE

DIPHENYL OXIDE

Products and Uses: An ingredient in perfumes and fragrances, particularly in soaps.

Precautions: Toxic by inhaling vapor.

Synonyms: CAS: 101-84-8 ✦ PHENYL ETHER ✦ DIPHENYL ETHER

DIPOTASSIUM PERSULFATE

Products and Uses: A defoaming agent for fresh citrus fruit juice.

Precautions: Moderately toxic in gross amounts. An irritant and allergen. FDA approves use at limited levels.

Synonyms: CAS: 7727-21-1 ✦ POTASSIUM PEROXYDISULFATE ✦ POTASSIUM PERSULFATE

DIPROPYLENE GLYCOL METHYL ETHER

Products and Uses: Ingredient in paints, inks, cosmetics, dyes, and pastes.

Precautions: Mildly toxic by swallowing and skin contact. A mild allergen and a skin and eye irritant.

Synonyms: CAS: 34590-940-8 ✦ DIPROPYLENE GLYCOL MONOMETHYL ETHER

DIPROPYLENE GLYCOL MONOSALICYLATE

Products and Uses: Found in sunscreening products and protective coatings as a UV (ultraviolet) light screening agent.

Precautions: Could cause allergic reaction.

Synonyms: DIPROPYLENE GLYCOL MONOESTER ✦ SALICYLIC ACID DIPROPYLENE GLYCOL MONOESTER

DIQUAT

Products and Uses: A herbicide and defoliant. Used as a plant growth regulator and suppressant.

Precautions: Poison by swallowing. A skin and eye irritant. FDA limits use.

Synonyms: CAS: 85-00-7 ✦ AQUACIDE ✦ DEIQUAT ✦ DEXTRONE ✦ DIQUAT DIBROMIDE ✦ FEGLOX ✦ PREEGLONE ✦ REGLON ✦ WEEDTRINE-D

DISODIUM EDTA

Products and Uses: An ingredient in salad dressings, margarines, mayonnaise, spreads, processed vegetables and fruits, soft drinks, and canned shellfish. As a preservative in foods. In medicine it is used as a chelating agent for treatment of lead poisoning.

Precautions: Moderately toxic by swallowing large amounts.

Synonyms: CAS: 139-33-3 ✦ ETHYLENEDIAMINE TETRAACETIC ACID ✦ DISODIUM SALT ✦ DISODIUM TETRACEMATE ✦ DISODIUM VERSENATE ✦ DISODIUM EDATHAMIL

DISODIUM INOSINATE

Products and Uses: A flavor enhancer for ham, meat (cured), poultry, and sausage.

Precautions: FDA approves use at levels to accomplish the intended effect. Moderately toxic.

Synonyms: CAS: 4691-65-0 ✦ DISODIUM IMP ✦ DISODIUM-5′-INOSINATE ✦ DISODIUM INOSINE-5′-MONOPHOSPHATE ✦ DISODIUM INOSINE-5′-PHOSPHATE

DISODIUM METHYLARSONATE

Products and Uses: A herbicide and defoliant (crabgrass killer).

Precautions: Toxic by swallowing and breathing.

Synonyms: CAS: 144-21-8 ✦ DMA ✦ DISODIUM METHANEARSONATE ✦ METHANEARSONIC ACID ✦ DISODIUM SALT

DISODIUM PYROPHOSPHATE

Products and Uses: An additive in biscuits, bologna, bologna (garlic), doughnuts, fish products (canned), frankfurters, hog carcasses, knockwurst, meat products, potatoes (processed), poultry, poultry food products, and Vienna wieners. As a sequestrant (binds constituents that affect the final product's appearance, flavor, or texture), an emulsifier (stabilizes and maintains mixes), and as a texturizer.

Precautions: GRAS (generally recognized as safe) by FDA when used within limitations.

Synonyms: CAS: 7758-16-9 ✦ DIPHOSPHORIC ACID, DISODIUM SALT ✦ DISODIUM DIHYDROGEN PYROPHOSPHATE ✦ DISODIUM DIPHOSPHATE ✦ SODIUM ACID PYROPHOSPHATE ✦ SODIUM PYROPHOSPHATE

DODECARBONIUM CHLORIDE

Products and Uses: Utilized as a biocide and a disinfectant.

Precautions: Toxic by swallowing.

Synonyms: NONE FOUND.

DODECENE

Products and Uses: Chemical used in flavors, perfumes, medicine, and oils.

Precautions: Irritant and narcotic in high concentrations.

Synonyms: CAS: 6842-15-5 ✦ 1-DODECENE ✦ DODECYLENE ✦ PROPENE TETRAMER ✦ TETRAPROPYLENE

DODECYL ALCOHOL

Products and Uses: Utilized as an odorant for detergents and perfumes; affects the smell of product.

Precautions: Mildly toxic by swallowing. A severe human skin irritant. Possible human carcinogen (may cause cancer).

Synonyms: CAS: 112-53-8 ✦ ALCOHOL C-12 ✦ ALFOL 12 ✦ CACHALOT ✦ n-DODECANOL ✦ 1-DODECANOL ✦ n-DODECYL ALCOHOL ✦ DUODECYL ALCOHOL ✦ LAURIC ALCOHOL ✦ LAURYL ALCOHOL

DODECYL BENZENE SODIUM SULFONATE

Products and Uses: Various uses in cosmetic ingredients and detergents.

Precautions: Moderately toxic by swallowing. A skin and severe eye irritant.

Synonyms: CAS: 25155-30-0 ✦ DETERGENT ALKYLATE ✦ BIO-SOFT ✦ DDBSA ✦ DODECYLBENZENESULFONIC ACID SODIUM SALT ✦ DODECYLBENZENESULPHONATE, SODIUM SALT ✦ SODIUM DODECYLBENZENESULFONATE ✦ SODIUM LAURYLBENZENESULFONATE ✦ SULFRAMIN ✦ ULTRAWET

DODECYLBENZYL MERCAPTAN

Products and Uses: A metal compound and an odorant chemical for cleaning and polishing metals. An odorant for warning of natural gas leaks.

Precautions: An irritant.

Synonyms: NONE FOUND.

DODECYL GALLATE

Products and Uses: An antioxidant (slows down the spoiling of fats due to oxidation) in cream cheese, fats, margarine, oil, and potatoes (instant mashed).

Precautions: FDA approves use within limitations.

Synonyms: CAS: 1166-52-5 ✦ LAURYL GALLATE ✦ NIPAGALLIN LA ✦ PROGALLIN LA

N-DODECYLSARCOSINE SODIUM SALT

Products and Uses: An antifogging agent and antistatic agent used in packaging materials.

Precautions: FDA approves use at moderate levels to accomplish the desired results.

Synonyms: NONE FOUND.

DODECYL TRIMETHYLAMMONIUM CHLORIDE

Products and Uses: Utilized as germicides and fungicides in fabric softeners and in mildew preventers.

Precautions: Could cause allergic reactions.

Synonyms: NONE FOUND.

ECHINACEA

Products and Uses: The roots and flowers of plants in the aster family. Considered by some to be an immunity booster and effective head cold treatment. Native Americans used it for more medicines than any other plant. Some controlled trials suggest that it can increase resistance to upper respiratory infections. It is thought to increase white blood cells; however, continued use decreases effectiveness. Commonly prescribed in the U.S. by physicians in the early 1900s. Today it is very frequently utilized in Europe. Products include tablets, liquids, and capsules.

Precautions: Reactions could occur to people allergic to sunflowers. It should not be used in incidences of impaired immunity, for example, tuberculosis, multiple sclerosis, and HIV.

Synonyms: PURPLE CONEFLOWER

ENZYMES

Products and Uses: Proteins that work as catalysts to enhance chemical reactions.

Precautions: Oral enzymes are broken down by digestion, as are other proteins, and are therefore of no special use in the body.

Synonyms: AMYLASE ✦ CARBOXYLASE ✦ CELLULASE ✦ CHOLINESTERASE ✦ CHYMOTRYPSIN ✦ INVERTASE ✦ LIPASES ✦ MALTASE ✦ PEPSIN ✦ PROTEASE ✦ RENNIN ✦ TRYPSIN ✦ RIBONUCLEASE ✦ UREASE* ZYMASE

EOSIN

Products and Uses: A dye and coloring agent in red ink, cosmetic products, lipstick, nail polish, and motor fuel coloring.

Precautions: FDA approves use in drugs and cosmetics except for use in eye area for eye makeup. Could cause an allergic reaction.

Synonyms: CAS: 15086-94-9 ✦ BROMEOSIN ✦ TETRABROMOFLUORESCEIN

EPHEDRINE

Products and Uses: A herbal stimulant found in asthma drugs and decongestants. Frequently promoted for weight reduction and energy booster. Commonly combined with caffeine.

Precautions: Can raise blood pressure, cause heart palpitations, stroke, memory loss, nerve damage, and psychosis. Some states and the FDA plan to restrict its use.

Synonyms: EPHEDRA ✦ MA HUANG ✦ EPITONIN ✦ PSEUDOEPHIDRINE

EPICHLOROHYDRIN

Products and Uses: It is a solvent in paints, varnishes, nail polishes, enamels, and lacquers.

Precautions: Toxic by breathing, swallowing, and skin absorption; strong irritant. A carcinogen (causes cancer).

Synonyms: CAS: 106-89-8 ✦ CHLOROPROPYLENE OXIDE ✦ CHLOROMETHYLOXIRANE

EPSOM SALTS

Products and Uses: Used medicinally, internally as a laxative and poison cathartic and for the treatment of cranial pressure and swelling. Topically, it is used as an anti-inflammatory. It also is used medically for anticonvulsive treatment.

Other uses include textile fireproofing, weighting silk, and dyeing. It is added to products including fertilizers, mineral waters, paper, and cosmetics, among others.

Precautions: Should not be consumed unless medically supervised as it is a toxic chemical.

Synonyms: MAGNESIUM SULFATE ✦ MAGNESIUM SULPHATE

ERGOSTEROL

Products and Uses: Found in yeast and mold; widely distributed in nature. When exposed to ultraviolet light it is converted to vitamin D, an antirachitic (promotes cure, prevents development of rickets).

Precautions: Overuse may be harmful due to its ability to catalyze calcium deposits in the bone.

Synonyms: PROVITAMINE D2

ERGOT

Products and Uses: A fungus. Source of many alkaloids and medicines for pharmaceuticals in appropriate doses. A vasoconstrictor.

Precautions: A poison by swallowing. Effects on the body by swallowing include nervousness, diarrhea, cyanosis (blue skin from lack of oxygen), excessive thirst, heart rhythm problems, gangrene, and circulatory changes.

Synonyms: CAS: 129-51-1 ✦ SECALE CORNUTUM ✦ RYE ERGOT ✦ RYE SCUM ✦ CRUDE ERGOT ✦ CORNOCENTIN ✦ ERGOTRATE

ERYTHORBIC ACID

Products and Uses: Found in bananas (frozen), beef (cured), cured meat food products, pork (cured), pork (fresh), and poultry as antioxidant (slows down spoiling), curing accelerator, and preservative.

Precautions: FDA states GRAS (generally recognized as safe) when used within limitations.

Synonyms: *d*-ARABOASCORBIC ACID

ERYTHRITE

Products and Uses: A coloring agent of crimson, peach, red, pink, or gray powder used by hobbyists in ceramic and glass work.

Precautions: Toxic by swallowing and breathing.

Synonyms: ERYTHRITOL ✦ COBALT BLOOM ✦ COBALT ARSENATE

ERYTHRITOL ANHYDRIDE

Products and Uses: Useful as a biocide or bacteriostat added to germicides and cleansers.

Precautions: Quite toxic.

Synonyms: CAS: 564-00-1 ✦ BUTADIENE DIOXIDE

ERYTHROSINE

Products and Uses: Utilized in food colors and stains.

Precautions: In large amounts it is moderately toxic by swallowing. A possible carcinogen (may cause cancer). A mutagen (changes inherited characteristics). FDA permits use only after certification. Use is then approved within limitations.

Synonyms: CAS: 16423-68-0 ✦ FD&C RED NO. 3 ✦ SODIUM SALT OF IODEOSIN

ESCULIN

Products and Uses: Derived from leaves and bark of horse chestnut tree. Used in skin lotions, creams, and liquids as a skin protectant.

Precautions: Could cause an allergic reaction in susceptible individuals. Harmless when used for intended purposes.

Synonyms: CAS: 531-75-9 ✦ ESCOSYL ✦ ESCULOSIDE ✦ 6,7-DIHYDROXYCOUMARIN ✦ 6-GLUCOSIDE

ESPARTO

Products and Uses: Used for carbon paper and other high quality papers. A wax that is used as a substitute for carnauba wax. It is derived from plant leaves of high cellulose content.

Precautions: Could cause allergic reaction.

Synonyms: NONE FOUND.

ESSENTIAL OIL

Products and Uses: In perfumes, air fresheners, sprays, deodorants, for example, oil of wintergreen, balsams, and oil of bitter almond. Previously obtained from leaves, flowers, bark, or stems of plants. They are mostly synthetic now. Used as odorants.

Precautions: Highly allergenic. Could cause eye or skin irritation upon contact. Possible respiratory effects when breathed.

Synonyms: SEE SPECIFIC EXAMPLES.

ESTERS OF FATTY ACIDS

Products and Uses: Various applications include adhesives, cosmetics, leather products, lubricants, and textile finishes.

Precautions: Harmless when used for intended purposes.

Synonyms: METHYL ESTER OF FATTY ACID ✦ BUTYL ESTER OF FATTY ACID ✦ PROPYL ESTER OF FATTY ACID ✦ GLYCERYL ESTER OF FATTY ACID ✦ POLYETHYLENE GLYCOL ESTER OF FATTY ACID

ETHANE

Products and Uses: Occurs in natural gas as a refrigerant or fuel.

Precautions: An asphyxiant gas. Severe fire risk if exposed to fire or open flame.

Synonyms: CAS: 74-84-0 ✦ DIMETHYL ✦ METHYLMETHANE

ETHANETHIOL

Products and Uses: It is an LPG (liquified petroleum gas) odorant. Used in adhesives. Tomato juice is reported to deodorize materials contaminated with this compound.

Precautions: Toxic by swallowing and breathing. Flammable. Dangerous fire risk. One of the most penetrating and persistent odors known (skunk).

Synonyms: CAS: 75-08-1 ✦ ETHYL SULFHYDRATE ✦ ETHYL MERCAPTAN

ETHANOLAMINE

Products and Uses: An additive for detergents in dry cleaning, wool treatments, paints, polishes, and sprays.

Precautions: Moderately toxic by swallowing and skin contact. A corrosive irritant to skin, eyes, nose, and throat. A mutagen (changes inherited characteristics).

Synonyms: CAS: 141-43-5 ✦ MEA ✦ MONOETHANOLAMINE ✦ COLAMINE ✦ 2-AMINOETHANOL ✦ 2-HYDROXYETHYLAMINE

ETHOHEXADIOL

Products and Uses: Applications include insect repellent, printing inks, grooming and hair care preparations. Used as an insecticide and for cosmetics.

Precautions: Moderately toxic by swallowing and skin contact. A skin and severe eye irritant.

Synonyms: CAS: 94-96-2 ✦ 6-12 INSECT REPELLENT ✦ ETHYL HEXANEDIOL ✦ OCTYLENE GLYCOL

ETHOXYQUIN

Products and Uses: Sprayed on apples after picking for preservative. It slows spoiling of fruit. Prevents mold growth and used as an antioxidant (slows down spoiling due to oxidation).

Precautions: Toxic by swallowing. Wash fruits well before consuming!

Synonyms: CAS: 91-53-2 ✦ SANTOQUINE ✦ SANTOFLEX ✦ SANTOQUIN ✦ STOP-SCALD

2-ETHOXETHYL-p-METHOXYCINNAMATE

Products and Uses: Frequently in suntan preparations as an ultraviolet (UV) absorber.

Precautions: Could cause allergic reactions.

Synonyms: NONE FOUND.

ETHYL ACETATE

Products and Uses: Found in coffee (decaffeinated), fruits, tea (decaffeinated), and vegetables as a coloring agent, flavoring agent, solvent, synthetic flavoring substance, and additive.

Precautions: Poison by breathing. In large amounts it is mildly toxic by swallowing. Effects by breathing are olfactory changes (ability of nose to smell), conjunctiva irritation (nose, throat, eyelid lining), and pulmonary changes (lungs and respiratory). A mutagen (changes inherited characteristics). Irritating to nose and throat surfaces, particularly the eyes, gums, and respiratory passages, and is also mildly narcotic. On repeated or prolonged exposures, it causes conjunctival irritation and corneal clouding. It can cause dermatitis. High concentrations have a narcotic effect and can cause congestion of the liver and kidneys. Chronic poisoning has been described as producing anemia, leucocytosis (transient increase in the white blood cell count), cloudy swelling, and fatty degeneration of the viscera (internal organs). FDA permits use within stated limits.

Synonyms: CAS: 141-78-6   ✦   ACETIC ETHER   ✦   ACETIDIN   ✦
ACETOXYETHANE   ✦   ETHYL ACETIC ESTER   ✦   ETHYL ETHANOATE   ✦
VINEGAR NAPHTHA

ETHYL ACRYLATE

Products and Uses: Applications include uses as a flavoring agent and for packaging materials.

Precautions: Confirmed carcinogen (causes cancer). In large amounts it is poison by swallowing and breathing. Moderately toxic by skin contact. Effects of breathing are eye, olfactory (nose and smelling), and pulmonary (lung) changes. A skin and eye irritant. Characterized in its terminal stages by dyspnea (shortness of breath), cyanosis (turning blue), and convulsive movements. It caused severe local irritation of the gastroenteric (intestinal) tract, and toxic degenerative changes of cardiac, hepatic (liver), renal (kidney), and splenic tissues were observed. No evidence of cumulative effects. A substance that migrates to food from packaging materials. FDA approves use at moderate levels to accomplish the intended effect.

Synonyms: CAS: 140-88-5   ✦   ACRYLIC ACID ETHYL ESTER   ✦
ETHOXYCARBONYLETHYLENE   ✦   ETHYL PROPENOATE   ✦
ETHYL-2-PROPENOATE

ETHYL ALCOHOL

Products and Uses: Used as an antimicrobial agent in detergents, cleaning preparations, and cosmetics. Used in gasohol, antifreeze, windshield defroster spray, and beverages.

Precautions: It is a confirmed carcinogen (causes cancer) by drinking of beverage alcohol. Moderately toxic to humans by swallowing. Mildly toxic by breathing and skin contact. Swallowing effects include sleep disorders, hallucinations, distorted perceptions, convulsions, motor activity changes, ataxia, coma, antipsychotic, headache, pulmonary changes, alteration in gastric secretion, nausea, or vomiting. Other effects include gastrointestinal changes, menstrual cycle changes and decreased body temperature. Can also cause glandular effects in humans. Reproductive effects (infertility or sterility or birth defects) by swallowing. FAS (fetal alcohol syndrome) effects on newborns as a result of the mother's drinking during pregnancy include changes in Apgar score, neonatal measures or effects, and drug dependence. A mutagen (changes inherited characteristics). An eye and skin irritant.

Synonyms: CAS: 64-17-5 ✦ ABSOLUTE ETHANOL ✦ ALCOHOL ✦ ALCOHOL, ANHYDROUS* ALCOHOL, DEHYDRATED* ALGRAIN ✦ ANHYDROL ✦ COLOGNE SPIRIT ✦ ETHANOL 200 PROOF ✦ ETHYL HYDRATE ✦ ETHYL HYDROXIDE ✦ FERMENTATION ALCOHOL ✦ GRAIN ALCOHOL ✦ METHYLCARBINOL ✦ MOLASSES ALCOHOL ✦ POTATO ALCOHOL

ETHYL ANTHRANILATE

Products and Uses: Frequently in fruit and floral flavors and aroma for beverages, ices, creams, baked products, and gelatins. Used as a perfume agent or flavoring agent. It is an ingredient that affects the taste or smell of final product.

Precautions: Moderately toxic by swallowing gross amounts. A skin irritant. FDA approves use at moderate levels to accomplish the intended effect.

Synonyms: CAS: 87-25-2 ✦ o-AMINOBENZOIC ACID, ETHYL ESTER ✦ ETHYL-o-AMINOBENZOATE

ETHYL BENZOATE

Products and Uses: Utilized in beverages, ice creams, ices, candies, bakery products, gum, gelatins, and liquors as artificial fruit (strawberry, cherry, grape, raspberry) and nut flavors.

Precautions: In large amounts it is moderately toxic by swallowing. Mildly toxic by skin contact. A skin and eye irritant.

Synonyms: CAS: 93-89-0 ✦ BENZOIC ACID ETHYL ESTER ✦ ETHYL BENZENECARBOXYLATE ✦ BENZOIC ETHER ✦ ESSENCE OF NIOBE

ETHYL BUTYRATE

Products and Uses: An additive in fruit, caramel, nut, beverage eggnog flavorings, ices, candies, bakery products, and puddings as flavoring; also used in perfumery. An ingredient that will affect the taste or smell of final product.

Precautions: In large amounts it is mildly toxic by swallowing. A skin irritant. FDA approves use at a moderate level.

Synonyms: CAS: 105-54-4 ✦ BUTANOIC ACID ETHYL ESTER ✦ BUTYRIC ETHER ✦ ETHYL BUTANOATE

ETHYL CAPROATE

Products and Uses: Common ingredient in artificial fruit essences. Useful as a flavoring agent.

Precautions: A skin irritant. FDA approves use in moderate amounts.

Synonyms: CAS: 123-66-0 ✦ ETHYL BUTYLACETATE ✦ ETHYL HEXANOATE

ETHYL CELLULOSE

Products and Uses: Frequently used as a confectionery, candy eggs (shell), food supplements in tablet form, gum, hot-glue adhesives, and printing inks. Useful as a binder and filler in dry vitamin preparations. A color fixer, a flavoring compound, and a coating (protective component for vitamin and mineral tablets).

Precautions: Harmless when used for intended purpose.

Synonyms: CAS: 9004-57-3 ✦ ETHYL ETHER OF CELLULOSE

ETHYL CINNAMATE

Products and Uses: An additive in perfumes, spicy and oriental soaps, colognes, and bath products. It is also in food spices, is a perfume fixative and food flavoring agent.

Precautions: Moderately toxic by swallowing large amounts. Could cause allergic effects.

Synonyms: CAS: 103-36-3 ✦ ETHYL-β-PHENYLACRYLATE ✦ ETHYL-3-PHENYLPROPENOATE ✦ CINNAMYLIC ETHER

ETHYL ENANTHATE

Products and Uses: Useful in liqueurs, soft drinks, cognac, berry, grape, cherry, apricot, and fruity-type soft drinks.

Precautions: Harmless when used for intended purposes.

Synonyms: ETHYL HEPTANOATE   ✦   COGNAC OIL   ✦   ARTIFICIAL COGNAC ESSENCE   ✦   OIL OF WINE

ETHYLENE DICHLORIDE

Products and Uses: Frequently in paint, varnish, gasoline, finish removers, soaps, scouring compounds, solvents, and fumigants.

Precautions: Toxic by swallowing, breathing, and skin absorption. Strong irritant to eyes and skin. A carcinogen (causes cancer).

Synonyms: CAS: 107-06-2   ✦   SYM-DICHLOROETHANE   ✦   ETHYLENE CHLORIDE   ✦   DUTCH OIL   ✦   1,2-DICHLOROETHANE

ETHYLENE GLYCOL

Products and Uses: An additive for cosmetics (up to 5%), printing inks, wood stains, adhesives, tobacco, ballpoint pen inks, stamp pad inks, and synthetic waxes; also for airplanes and runways. Extremely varied uses include as coolant, antifreeze, deicer, and moisturizer.

Precautions: Human poison by swallowing. Lethal dose for humans reported to be 100 ml. Effects by breathing and swallowing are eye watering, anesthesia, headache, cough, nausea, vomiting, lung, kidney, and liver changes. If swallowed it causes central nervous system stimulation followed by depression. Ultimately it causes potential lethal kidney damage. Very toxic in particulate form upon breathing. A mutagen (changes inherited characteristics). A skin, eye, nose, and throat irritant.

Synonyms: CAS: 107-21-1   ✦   ETHYLENE ALCOHOL   ✦   GLYCOL   ✦   1,2-ETHANEDIOL

ETHYLENE GLYCOL METHYL ETHER

Products and Uses: A component in confectioneries and food supplements in tablet form. Applications include uses in gum and poultry as a binder, color fixer, and extender.

Precautions: Moderately toxic to humans by swallowing large amounts. Effects by breathing are change in motor activity, tremors, and convulsions. A skin and eye irritant. When used under conditions which do not require the application of heat, this material probably presents little hazard to health.

Synonyms: CAS: 109-86-4 ✦ GLYCOL ETHER EM ✦ GLYCOLMETHYL ETHER ✦ GLYCOL MONOMETHYL ETHER ✦ 2-METHOXYETHANOL ✦ METHOXYHYDROXYETHANE ✦ METHYL CELLOSOLVE ✦ METHYL ETHOXOL ✦ METHYL GLYCOL ✦ METHYL OXITOL

ETHYLENE GLYCOL DIACETATE

Products and Uses: Commonly included in lacquers, printing inks, and perfumes as a solvent and fixative.

Precautions: Mildly toxic by swallowing and skin contact. An eye irritant.

Synonyms: CAS: 111-55-7 ✦ GLYCOL DIACETATE ✦ ETHLENE ACETATE ✦ ETHYLENE GLYCOL ACETATE

ETHYLENE GLYCOL MONOBUTYL ETHER

Products and Uses: Commonly used in spray lacquers, quick-drying lacquers, varnishes, enamels, dry-cleaning compounds, and varnish removers. As a solvent and in soap emulsifier.

Precautions: Poison by swallowing and skin contact. Moderately toxic by breathing. Effects by breathing are nausea, vomiting, headache, and nose tumors. A skin irritant.

Synonyms: CAS: 111-76-2 ✦ 2-BUTOXYETHANOL ✦ BUTYL CELLOSOLVE

ETHYLENE GLYCOL MONOETHYL ETHER

Products and Uses: Useful in varnish remover, cleaning solutions, anti-icing additive for fuels, and as a solvent and antifreeze.

Precautions: Moderately toxic by swallowing and skin contact. Mildly toxic by breathing. An eye and skin irritant.

Synonyms: CAS: 110-80-5 ✦ 2-ETHOXYETHANOL ✦ CELLOSOVE SOLVENT

ETHYLENE GLYCOL MONOMETHYL ETHER

Products and Uses: An ingredient in lacquers, enamels, varnishes, perfumes, wood stains, sealing, moisture-proof cellophane, and jet-fuel deicing fluids. Used as a solvent, fixative, and additive.

Precautions: Toxic by swallowing and breathing.

Synonyms: CAS: 109-86-4 ✦ 2-METHOXYMETHANOL ✦ METHYL CELLOSOLVE

ETHYLENEIMINE

Products and Uses: In adhesives, fuel oil, lubricants, and germicides for microbial control, as additive, and an intermediate (used to make other chemicals).

Precautions: Corrosive. Absorbed by skin and causes tumors. A carcinogen (causes cancer). Could cause allergic skin reaction.

Synonyms: CAS: 151-56-4 ✦ AZRIDINE ✦ ETHYLENIMINE

ETHYL FORMATE

Products and Uses: It occurs naturally in apples and coffee extract. Used for lemonades, baked goods, candy (hard), candy (soft), chewing gum, fillings, frozen dairy desserts, gelatins, puddings, raisins, and Zante currents. Additive in fumigants and larvicides as a flavoring agent, fruit essence, and insecticide.

Precautions: Moderately toxic by swallowing large amounts. Mildly toxic by skin contact and breathing. A powerful breathing irritant in humans. A skin and eye irritant. Questionable carcinogen (may cause cancer). Highly flammable liquid. FDA approves use within limits.

Synonyms: CAS: 109-94-4 ✦ AREGINAL ✦ ETHYL FORMIC ESTER ✦ ETHYL METHANOATE ✦ FORMIC ACID, ETHYL ESTER ✦ FORMIC ETHER

ETHYL LAURATE

Products and Uses: Commonly utilized in fruit, nut, liquor, spice, cheese flavorings, beverage flavorings, ices, creams, bakery products, and gum as a solvent and for food flavoring.

Precautions: Harmless when used for intended purposes.

Synonyms: CAS: 106-33-2 ✦ ETHYL DODECANOATE

ETHYL MALTOL

Products and Uses: An additive in chocolates, desserts, and wines, used as a flavoring agent and processing aid.

Precautions: In gross amounts it is moderately toxic by swallowing. FDA approves use at moderate levels to accomplish the intended effects.

Synonyms: CAS: 4940-11-8 ✦ 2-ETHYL-3-HYDROXY-4H-PYRAN-4-ONE ✦ 2-ETHYL PYROMECONIC ACID ✦ 3-HYDROXY-2-ETHYL-4-PYRONE

ETHYLMERCURIC PHOSPHATE

Products and Uses: Utilized for seeds and wood as a fungicide and preservative.

Precautions: Poison by swallowing.

Synonyms: CAS: 2440-24-1 ✦ LIGNASAN FUNGICIDE ✦ CERESAN ✦ GRANOSAN

ETHYL METHYLPHENYLGLYCIDATE

Products and Uses: Frequently used in beverages, candy, and ice cream as a flavoring agent.

Precautions: Mildly toxic by swallowing large amounts. FDA approves use at moderate levels to accomplish the desired results.

Synonyms: CAS: 77-83-8 ✦ C-16 ALDEHYDE ✦ EMPG ✦ α-β-EPOXY-β-METHYLHYDROCINNAMIC ACID, ETHYL ESTER ✦ ETHYL α,β-EPOXY-β-METHYLHYDROCINNAMATE ✦ ETHYL 2,3-EPOXY-3-METHYL-3-PHENYLPROPIONATE ✦ ETHYL ESTER of 2,3-EPOXY-3-PHENYLBUTANOIC ACID ✦ FRAESEOL ✦ 3-METHYL-3-PHENYLGLYCIDIC ACID ETHYL ESTER ✦ STRAWBERRY ALDEHYDE

ETHYL NONANOATE

Products and Uses: Common ingredient in fruit and rum-flavored beverages, beverages (alcoholic), candy, ice cream, bakery products, gelatins, puddings, desserts, and icings as a synthetic flavoring agent.

Precautions: Mildly toxic by swallowing gross amounts. A skin irritant. FDA approves use at moderate levels to accomplish the intended effect.

Synonyms: CAS: 123-29-5 ✦ ETHYL NONYLATE ✦ ETHYL PELARGONATE ✦ NONANOIC ACID, ETHYL ESTER ✦ WINE ETHER

ETHYL PELARGONATE

Products and Uses: In alcoholic beverages (cognac essence), perfumes, and flame retardants. For flavorings and perfumery.

Precautions: An irritant to eyes and lungs.

Synonyms: CAS: 2524-04-1 ✦ ETHYL NONANOATE ✦ WINE ETHER ✦ ETHYL PCT

ETHYL PHENYLACETATE

Products and Uses: Popular ingredient in honey, butter, fruit flavoring, beverage flavoring, creams, ices, candy, and bakery products for food product flavoring, perfume, aromas, and scents.

Precautions: In large amounts it is moderately toxic by swallowing. FDA approves use at moderate levels to accomplish the intended effect.

Synonyms: CAS: 101-97-3 ✦ BENZENEACETIC ACID, ETHYL ESTER ✦ ETHYL BENZENEACETATE ✦ ETHYL PHENACETATE ✦ ETHYL-2-PHENYLETHANOATE ✦ ETHYL-α-TOLUATE ✦ PHENYLACETIC ACID, ETHYL ESTER ✦ α-TOLUIC ACID, ETHYL ESTER

ETHYLPHOSPHORIC ACID

Products and Uses: Chemical useful for metal items as rust remover.

Precautions: An irritant. Follow directions carefully when using products that contain this chemical.

Synonyms: NONE FOUND.

ETHYL SILICATE

Products and Uses: In coatings, paints, lacquers, and bondings. For weatherproof mortar, cements, bricks, and heat resistant paints.

Precautions: Strong irritant to eyes, nose, and throat.

Synonyms: CAS: 78-10-4 ✦ TETRAETHYL ✦ ORTHOSILICATE

EUCALYPTUS OIL

Products and Uses: Used in bakery products, beverages (alcoholic), beverages (nonalcoholic), confections, and ice cream. Also utilized in pharmaceuticals and antibacterials. It is considered a flavoring agent and odorant.

Precautions: A poison by swallowing large amounts. Effects by swallowing are eye spasms, sleepiness, and respiratory depression. A skin irritant. FDA approves use at moderate levels to accomplish the desired results.

Synonyms: CAS: 8000-48-4 ✦ DINKUM OIL ✦ EUKALYPTUS OEL ✦ OIL OF EUCALYPTUS

EUGENOL

Products and Uses: Various applications include baked goods, beverages (nonalcoholic), candy, chewing gum, condiments, confections, gelatin desserts, ice cream, meat, puddings, and spice oils. Used in medications as a flavoring agent and in perfumery to replace oil of cloves. It is also an insect attractant. Used as a dental analgesic (pain killer).

Precautions: In large amounts it is moderately toxic by swallowing. A mutagen (changes inherited characteristics). A skin irritant. GRAS (generally recognized as safe).

Synonyms: CAS: 97-53-0 ✦ 4-ALLYLGUAIACOL ✦ 4-ALLYL-1-HYDROXY-2-METHOXYBENZENE ✦ 4-ALLYL-2-METHOXYPHENOL ✦ CARYOPHYLLIC ACID ✦ EUGENIC ACID ✦ SYNTHETIC EUGENOL

F

FARNESOL

Products and Uses: An additive in beverages, ice creams, ices, bakery products, and puddings. It occurs naturally in anise, flowers, oils, and roses. Used as apricot, banana, melon, and berry flavoring.

Precautions: FDA approves use in limited amounts. Mildly toxic by swallowing gross amounts. A mutagen (changes inherited characteristics)

Synonyms: CAS: 4602-84-0 ✦ FARNESYL ALCOHOL ✦
3,7,11-TRIMETHYL-2,6,10-DODECATRIEN-1-OL

FATTY ACIDS

Products and Uses: A basic ingredient in soaps, detergents, bubble baths, lipstick, candles, salad oils, shortenings, and cosmetics. A component in the manufacture of food-grade additives, as a defoaming agent, lubricant, emulsifier, and binder.

Precautions: Harmless when used for intended purpose.

Synonyms: CAPRIC ACID ✦ CAPRYLIC ACID ✦ LAURIC ACID ✦ MYRISTIC ACID ✦ OLEIC ACID ✦ PALMITIC ACID ✦ STEARIC ACID

FD&C BLUE No. 1

Products and Uses: Used in baked goods, candy, confections, food, drugs, toothpaste, hair coloring, and cosmetics.

Precautions: A possible carcinogen (causes cancer). Could cause allergic reaction. FDA approves use at moderate levels to accomplish the desired results.

Synonyms: CAS: 38444-45-9 ✦ ACID SKY BLUE A ✦ AIZEN FOOD BLUE No. 2 ✦ BRILLIANT BLUE FCD No. 1 ✦ BRILLIANT BLUE FCF ✦ CANACERT

BRILLIANT BLUE FCF ✦ COSMETIC BLUE LAKE ✦ D&C BLUE No. 4 ✦ DOLKWAL BRILLIANT BLUE ✦ FOOD BLUE 2 ✦ FOOD BLUE DYE NO. 1 ✦ HEXACOL BRILLIANT BLUE A ✦ INTRACID PURE BLUE L

FD&C RED No. 3

Products and Uses: A coloring agent for candy, cherries, confections, toothpaste, cereals, gelatins, and printing inks.

Precautions: In gross amounts it is moderately toxic by swallowing. A possible carcinogen.

Synonyms: CAS: 16423-68-0 ✦ AIZEN ERYTHROSINE ✦ CALCOCID ERYTHROSINE N ✦ CANACERT ERYTHROSINE BS ✦ 9-(o-CARBOXYPHENYL)-6-HYDROXY-2,4,5,7-TETRAIODO-3-ISOXANTHONE ✦ CILEFA PINK B ✦ D&C RED No. 3 ✦ DOLKWAL ERYTHROSINE ✦ DYE FD&C RED No. 3 ✦ ERYTHROSIN ✦ MAPLE ERYTHROSINE ✦ 2′4′,5′,7′-TETRAIODOFLUORESCEIN, DISODIUM SALT ✦ TETRAIODOFLUORESCEIN SODIUM SALT ✦ USACERT RED No. 3

FD&C YELLOW No. 5

Products and Uses: A color additive in butter, cheese, ice cream, and ink.

Precautions: Suspected to be the cause of allergy, especially in aspirin-sensitive individuals. In gross amounts it is mildly toxic by swallowing. Effects on the body by swallowing are paresthesia (abnormal sensation of burning and tingling) and changes in teeth and supporting structures.

Synonyms: CAS: 1934-21-0 ✦ ACID LEATHER YELLOW T ✦ AIREDALE YELLOW T ✦ AIZEN TARTRAZINE ✦ ATUL TARTRAZINE ✦ BUCACID TARTRAZINE ✦ D&C YELLOW No. 5 ✦ DOLKWAL TARTRAZINE ✦ EGG YELLOW A ✦ EUROCERT TARTRAZINE ✦ FOOD YELLOW No. 4 ✦ HEXACOL TARTRAZINE ✦ HYDRAZINE YELLOW ✦ KARO TARTRAZINE ✦ LAKE YELLOW ✦ MAPLE TARTRAZOL YELLOW ✦ TARTAR YELLOW ✦ TARTRAZINE ✦ TARTRAZOL YELLOW ✦ TRISODIUM-3-CARBOXY-5-HYDROXY-1-p-SULFOPHENYL-4-p-SULFOPHENYLAZOPYRAZOLE ✦ VONDACID TARTRAZINE ✦ WOOL YELLOW ✦ YELLOW LAKE 69

FELDSPAR

Products and Uses: As a filler, grit, and glazing material in soaps, cements, tarred roofing materials, pottery, enamelware, ceramic ware, glass, and fertilizer.

Precautions: Toxic if consumed as a fine-ground powder.

Synonyms: POTASSIUM ALUMINOSILICATE

FENNEL OIL

Products and Uses: Useful in seasoning for sausage, beverages, fruit products, bakery products, meats, root beer, and condiments as a food and beverage flavoring or seasoning.

Precautions: Moderately toxic by swallowing gross amounts. A mutagen (changes inherited characteristics). A severe skin irritant. Could cause allergic reaction. GRAS (generally recognized as safe) when used at moderate levels to accomplish the desired results.

Synonyms: CAS: 8006-84-6 ✦ BITTER FENNEL OIL ✦ FENCHEL OEL (GERMAN) ✦ OIL OF FENNEL

FENURON

Products and Uses: Chemical herbicide used as a weed and brush killer.

Precautions: Moderately toxic by swallowing.

Synonyms: CAS: 101-42-8 ✦ FENULON ✦ DIBAR ✦ PUD ✦ FENIDIN ✦ 1,1-DIMETHYL-3-PHENYLUREA

FERRIC CHLORIDE, ANHYDROUS

Products and Uses: Utilized in pigments, inks, astringents, styptics, dyes, and inks as a disinfectant, sewage deodorizer, and water purifier.

Precautions: Toxic by swallowing. A strong irritant to skin and tissue.

Synonyms: CAS: 7705-08-0 ✦ FERRIC TRICHLORIDE ✦ FERRIC PERCHLORIDE ✦ IRON CHLORIDE ✦ IRON TRICHLORIDE ✦ IRON PERCHLORIDE

FERRIC PHOSPHATE

Products and Uses: A component in egg substitutes (frozen), pasta products, and rice products. It is considered a nutritional supplement.

Precautions: GRAS (generally recognized as safe).

Synonyms: CAS: 10045-86-0 ✦ FERRIC ORTHOPHOSPHATE ✦ IRON PHOSPHATE

FERRIC STEARATE

Products and Uses: An additive in varnishes and photocopy machine chemicals; useful as a drier.

Precautions: Could cause allergic reaction to skin.

Synonyms: IRON STEARATE

FERRIC SULFATE

Products and Uses: Found in personal care, hair coloring, and dye products as a deodorant, astringent, disinfectant; also in textile printing inks.

Precautions: GRAS (generally recognized as safe).

Synonyms: CAS: 10028-22-5 ✦ DIIRON TRISULFATE ✦ IRON PERSULFATE ✦ IRON SESQUISULFATE ✦ IRON SULFATE (2:3) ✦ IRON(III) SULFATE ✦ IRON TERSULFATE ✦ SULFURIC ACID, IRON (3⁺) SALT (3:2)

FERROUS FUMARATE

Products and Uses: Used as a dietary supplement and nutritional supplement in cereals and waffles.

Precautions: In gross amounts it is moderately toxic by swallowing. GRAS (generally recognized as safe). Limited as a source of iron for special dietary purposes.

Synonyms: CAS: 141-01-5 ✦ CPIRON ✦ ERCO-FER ✦ ERCOFERRO ✦ FEOSTAT ✦ FEROTON ✦ FERROFUME ✦ FERRONAT ✦ FERRONE ✦ FERROTEMP ✦ FERRUM ✦ FERSAMAL ✦ FIRON ✦ FUMAFER ✦ FUMAR-F ✦ FUMIRON ✦ GALFER ✦ HEMOTON ✦ IRCON ✦ IRON FUMARATE ✦ METERFER ✦ METERFOLIC ✦ ONE-IRON ✦ PALAFER ✦ TOLERON ✦ TOLFERAIN ✦ TOLIFER

FERROUS GLUCONATE

Products and Uses: Useful for olives (ripe) and vitamin pills as a color additive, dietary, and nutrient supplement.

Precautions: Moderately toxic by swallowing gross amounts. Effects on the body by ingesting are diarrhea, nausea, and vomiting. GRAS (generally recognized as safe).

Synonyms: CAS: 299-29-6 ✦ FERGON ✦ FERGON PREPARATIONS ✦ FERLUCON ✦ FERRONICUM ✦ GLUCO-FERRUM ✦ IROMIN ✦ IRON GLUCONATE ✦ IROX (GADOR) ✦ NIONATE ✦ RAY-GLUCIRON

FERROUS OXALATE

Products and Uses: In pigments and photographic developer for coloration.

Precautions: Toxic.

Synonyms: CAS: 516-03-0 ✦ IRON OXALATE

FERROUS SULFATE

Products and Uses: An additive in wood preservative and wine; also used as a dietary supplement and nutrient.

Precautions: It is a human poison by swallowing large amounts. Effects on the body by swallowing are aggression, somnolence, brain recording changes, diarrhea, nausea, vomiting, bleeding from the stomach, and coma. A mutagen (changes inherited characteristics). GRAS (generally recognized as safe) when used at moderate levels.

Synonyms: CAS: 7720-78-7 ✦ COPPERAS ✦ DURETTER ✦ DUROFERON ✦ EXSICCATED FERROUS SULPHATE ✦ FEOSOL ✦ FEOSPAN ✦ FER-IN-SOL ✦ FERRO-GRADUMET ✦ FERROSULFATE ✦ FERRO-THERON ✦ FERSOLATE ✦ GREEN VITRIOL ✦ IRON MONOSULFATE ✦ IRON PROTOSULFATE ✦ IRON(II) SULFATE (1:1) ✦ IRON VITRIOL ✦ IROSPAN ✦ IROSUL ✦ SLOW-FE ✦ SULFERROUS ✦ SULFURIC ACID, IRON(2$^+$) SALT (1:1)

FEVERFEW

Products and Uses: The leaves of the aster flower. Commonly used in Europe for the relief of migraine headaches and vomiting symptoms.

Precautions: Mouth ulceration and inflammation has occurred in patients who consumed and chewed fresh leaves.

Synonyms: TANACETUM PARTHENIUM ✦ CHRYSANTHEMUM PARTHENIUM

FICIN

Products and Uses: Derived from the *Ficus* fig tree. Used in beer, cereals (pre-cooked), meat (raw cuts), poultry, and wine, for chillproofing of beer, coagulation of milk, as an enzyme, processing aid, and as a meat tenderizing agent.

Precautions: Mildly toxic by swallowing gross amounts.

Synonyms: CAS: 9001-33-6 ✦ DEBRICIN ✦ FICUS PROTEASE ✦ FICUS PROTEINASE ✦ HIGUEROXYL DELABARRE ✦ TL 367

FIR NEEDLE OIL, CANADIAN TYPE

Products and Uses: Obtained from the needles and twigs of balsams for perfumes and foods, as a fragrance, odorant, and flavoring.

Precautions: FDA approves use at moderate levels to accomplish the desired results. Could cause allergic reactions.

Synonyms: BALSAM FIR OIL

FISH GLUE

Products and Uses: Applications include gummed tape, cartons, blueprint paper, and printing plates. A ton of skins yields about 50 gallons of liquid glue and adhesive.

Precautions: Harmless when used for intended purposes. Susceptible individuals have experienced skin irritation.

Synonyms: NONE FOUND.

FLUORIDE

Products and Uses: A tooth cavity preventative found in toothpastes, teeth rinses, and dentifrices.

Precautions: Harmless when used at rate of 1 ppm in drinking water for the purpose of reducing tooth decay. Excessive doses can cause nausea, vomiting, diarrhea, and cramps. Can aggravate asthma and cause severe bone changes, making movement painful. Irritant to the eyes, skin, nose, and throat.

Synonyms: FLUOSILICIC ACID ✦ SODIUM SILICOFLUORIDE ✦ SODIUM FLUORIDE

FLUOSILICIC ACID

Products and Uses: Commonly used in cement, wood preservative, and disinfectant for hardening and sterilizing.

Precautions: Extremely corrosive by skin contact and breathing.

Synonyms: CAS: 16961-83-4 ✦ HYDROFLUOSILICIC ACID ✦ FLUOROSILIC ACID ✦ HEXAFLUORISILIC ACID ✦ HYDROGEN HEXAFLUORISILICATE

FOLIC ACID

Products and Uses: Found in green vegetables, liver, kidney, nuts, and orange juice. Also added to cosmetics; a dietary supplement and a skin softener. Some believe that folic acid can help prevent heart disease and strokes. Stroke and heart attack victims have high levels of homocysteine. Lack of folic acid results in high homocysteine levels. High levels of this amino acid can contribute to neurological birth defects such as spina bifida and may cause damage to arteries.

Precautions: A member of the vitamin B complex. FDA approves use within limitations. Recommended dietary allowance is 180 mcg a day.

Synonyms: CAS: 59-30-3 ✦ *l*-N-(p-(((-2-AMINO-4-HYDROXY-6-PTERIDINYL) METHYL)AMINO)BENZOYL)GLUTAMIC ACID ✦ FOLACIN ✦ FOLATE ✦ FOLCYSTEINE ✦ PTEGLU ✦ PTEROYLGLUTAMIC ACID ✦ PTEROYL-*l*-GLUTAMIC ACID ✦ PTEROYLMONOGLUTAMIC ACID ✦ PTEROYL-*l*-MONOGLUTAMIC ACID ✦ VITAMIN Bc ✦ VITAMIN M

FOLPET

Products and Uses: Ingredient in paints and enamels. Its purpose is as a fungicide and bactericide.

Precautions: Moderately toxic by swallowing.

Synonyms: CAS: 133-07-3 ✦ PHALTAN ✦ FOLPAN ✦ THIOPHAL ✦ ORTHOPHALTAN

FORMALDEHYDE

Products and Uses: Commonly used in glues, air fresheners, antiperspirants, dry-cleaning solvents, fingernail polish, hair spray, laundry spray starch, perfumes, after-shave lotions, preservatives, cottonseed, packaging materials, fabric durable press treatment, particle board, and plywood. As a deodorizer, solvent, preservative, disinfectant, and adhesive, among others.

Precautions: A confirmed carcinogen (causes cancer). A poison by swallowing. Effects on the body by swallowing are lacrimation (causes eye watering), olfactory changes (relating to the ability to smell), aggression, and pulmonary (lung) damage. A human mutagen (changes inherited characteristics). A skin and eye irritant. If swallowed, it causes violent vomiting and diarrhea, which can lead to collapse. Frequent or prolonged exposure can cause hypersensitivity leading to contact dermatitis (skin irritation), possibly of an eczematoid (resembling eczema) nature. An air concentration of 20 ppm is quickly irritating to eyes. Outgassing of chemical from building materials in homes and trailers is a problem. Possibly one of the factors in "sick building" syndrome.

Synonyms: CAS: 50-00-0 ✦ BFV FANNOFORM ✦ FORMALDEHYDE, SOLUTION(DOT) ✦ FORMALIN ✦ FORMALIN 40 ✦ FORMALIN (DOT) ✦ FORMALITH ✦ FORMIC ALDEHYDE ✦ FORMOL ✦ FYDE ✦ HOCH ✦ IVALON ✦ KARSAN ✦ LYSOFORM ✦ METHANAL ✦ METHYL ALDEHYDE ✦ METHYLENE GLYCOL ✦ METHYLENE OXIDE ✦ MORBOCID ✦ OXOMETHANE ✦ OXYMETHYLENE ✦ PARAFORM ✦ POLYOXYMETHYLENE GLYCOLS ✦ SUPERLYSOFORM

FORMIC ACID

Products and Uses: A component in food packaging, synthetic flavorings, hair products, perfumes, lacquers, and refrigerants. As a fumigant, brewing antiseptic, solvent, and plasticizer.

Precautions: Moderately toxic by swallowing and mildly toxic by breathing. Corrosive. A skin and severe eye irritant. A substance migrating to food from packaging materials. GRAS (generally recognized as safe) as an indirect additive.

Synonyms: CAS: 64-18-6 ✦ AMINIC ACID ✦ FORMYLIC ACID ✦ HYDROGEN CARBOXYLIC ACID ✦ METHANOIC ACID

FUMARIC ACID

Products and Uses: Versatile chemical utilized in printing inks, beverage mixes (dry), candy, desserts, pie fillings, poultry, and wine. As an acidifier, tartness agent, curing accelerator, and flavoring agent.

Precautions: In large amounts it is mildly toxic by swallowing. Toxic by skin contact. A skin and eye irritant. FDA approves use at moderate levels to accomplish the intended effect when used within limits.

Synonyms: CAS: 110-17-8 ✦ ALLOMALEIC ACID ✦ BOLETIC ACID ✦ trans-BUTENEDIOIC ACID ✦ (E)-BUTENEDIOIC ACID ✦ trans-1,2-ETHYLENEDICARBOXYLIC ACID ✦ (E)1,2-ETHYLENEDICARBOXYLIC ACID ✦ LICHENIC ACID

FURCELLERAN GUM

Products and Uses: Additive in flans, jams, jellies, meat products (gelled), puddings (milk), and toothpastes. Harmless when used for intended purposes. Used as an emulsifier, stabilizer, thickening agent, and gelling agent.

Precautions: In great amounts it is moderately toxic by swallowing.

Synonyms: CAS: 9000-21-9 ✦ BURTONITE 44

FURFURAL

Products and Uses: Varied applications include in shoe dyes; also as a weed killer, and fungicide. Also an ingredient in caramel, butterscotch, fruit, and molasses flavors. In nut flavors for beverages and ice cream dessert syrups. Used as a herbicide, leather preservative, and flavoring.

Precautions: Poison by swallowing large amounts. Moderately toxic by breathing, and skin contact. Also an eye irritant. The liquid is dangerous to the eyes. The vapor is irritating to nose and throat and is a central nervous system poison. GRAS (generally recognized as safe) when used in moderate amounts.

Synonyms: CAS: 98-01-1 ✦ ARTIFICIAL ANT OIL ✦ FURAL ✦ FURALE ✦ 2-FURALDEHYDE ✦ 2-FURANALDEHYDE ✦ 2-FURANCARBONAL ✦ 2-FURANCARBOXALDEHYDE ✦ 2-FURFURAL ✦ FURFURALDEHYDE ✦ FURFUROL ✦ FURFUROLE ✦ FUROLE ✦ α-FUROLE ✦ 2-FURYL-METHANAL ✦ PYROMUCIC ALDEHYDE

FUSEL OIL

Products and Uses: A component in explosives, varnishes, lacquers, fats, oils, waxes, and perfumery. Also used in liquor, wine, whiskey, as a food and beverage seasoning, and as a solvent, flavoring, and gelatinizing agent.

Precautions: May contain carcinogens. Toxic by swallowing and breathing. FDA states use should be at moderate level to accomplish the desired results.

Synonyms: CAS: 8013-75-0 ✦ FUSEL OIL, REFINED

GALLIC ACID

Products and Uses: An ingredient in inks, dyes, tanning agents, photography chemicals, and pharmaceuticals; useful for marker colors, leather coloring, and dyeing.

Precautions: Mildly toxic by swallowing. Could cause allergic reaction.

Synonyms: CAS: 149-91-7 ✦ 3,4,5-TRIHYDROXYBENZOIC ACID

GARLIC

Products and Uses: Used as a food and medicine since before the time of Christ. Universally acclaimed for reducing cholesteral and triglycerides, reducing blood pressure and a preventative of cancers, colds and respiratory infections. Considered to have anti-inflammatory, antifungul, antioxidant and antibacterial properties. The typical dose is 900 mg of powdered garlic standardized to 0.6% allicin per 100 mg or 5.4 mg allicin to lower cholesterol.

Precautions: Infrequently found to be an allergen.

Synonyms: NONE FOUND.

GENTIAN VIOLET

Products and Uses: Derived from an herb. Used to treat poultry for fungus disease and added to poultry feed to prevent fungus. Also has been used to treat humans for burn wounds, fungus infections of skin, and internally for parasite infections

Precautions: A carcinogen.

Synonyms: CRYSTAL VIOLET

GERANIUM OIL, ALGERIAN TYPE

Products and Uses: Used in bakery products, beverages (nonalcoholic), chewing gum, confections, gelatin desserts, ice cream, and puddings. Used as a flavoring agent.

Precautions: A skin irritant. GRAS (generally recognized as safe).

Synonyms: CAS: 8000-46-2　✦　GERANIUM OIL　✦　OIL OF GERANIUM　✦　OIL OF PELARGONIUM　✦　OIL OF ROSE GERANIUM　✦　OIL ROSE GERANIUM ALGERIAN　✦　PELARGONIUM OIL　✦　ROSE GERANIUM OIL ALGERIAN

GERANYL BUTYRATE

Products and Uses: Used in berry and fruit flavorings for beverages, desserts, candy, and bakery products. Typically found in rose-scented perfumes and soaps as a flavoring agent and fragrance.

Precautions: Harmless when used for intended purposes.

Synonyms: 3,7-DIMETHYL-2,6-OCTADIENE-1-YL BUTYRATEESTER　✦　(E)-3-METHYLBUTYRIC ACID-3,7-DIMETHYL-2,6-OCTADIENYL ESTER

GHATTI GUM

Products and Uses: Utilized in butter and butterscotch flavorings. It is an emulsifier (stabilizes mixes to aid in suspension of liquids) for oils, fats, and waxes. Used as a thickener and flavoring.

Precautions: Could cause allergic reaction.

Synonyms: NONE FOUND.

GINSENG

Products and Uses: The root of an herb used by the Chinese for medicinal purposes for 2000 years. Currently a popular item in the U.S. This supplement is promoted as being beneficial for almost all human ailments, including stress, depression, anemia, diabetes, menopause, and so on. There are no long-term studies to support these claims

Precautions: Use at typical dosage appears to be harmless. Those with high blood pressure and pregnant patients should avoid use.

Synonyms: PANAX GINSENG　✦　GINSENOSIDE

GLUCONO Δ-LACTONE

Products and Uses: Used in dessert mixes, frankfurters, genoa salamis, meat food products, and sausages as an acidifier (to make tart), binder (to hold ingredients together), curing agent, leavening agent, pickling agent, sequestrant (affects the final product's appearance or taste).

Precautions: GRAS (generally recognized as safe) by FDA when used within limits for intended purposes.

Synonyms: CAS: 90-80-2

d-GLUCOSE

Products and Uses: An ingredient in bakery products, baby foods, confections, ham (chopped or processed), hamburger, ice cream, luncheon meat, meat loaf, poultry, and sausage. Used in brewing and wine making and as a formulation aid, humectant (moisturizer), sweetener (nutritive), texturizing agent (improves feel or texture).

Precautions: Safe for the general population in amounts usually consumed.

Synonyms: CAS: 50-99-7 ✦ CARTOSE ✦ CERELOSE ✦ CORN SUGAR ✦ DEXTROPUR ✦ DEXTROSE ✦ DEXTROSE, ANHYDROUS ✦ DEXTROSOL ✦ GLUCOLIN ✦ GLUCOSE ✦ *d*-GLUCOSE, ANHYDROUS ✦ GLUCOSE LIQUID ✦ GRAPE SUGAR ✦ SIRUP

l-GLUTAMIC ACID

Products and Uses: Used in foods as a flavor enhancer, in tobacco as a taste enhancer, in cosmetics and toiletries as a preservative. Also considered a dietary supplement, nutrient, antioxidant (slows down spoiling), and a salt substitute.

Precautions: In excessive amounts the effect on the human body by swallowing are headache and nausea or vomiting. GRAS (generally recognized as safe) by the FDA when used at moderate levels to accomplish the desired results.

Synonyms: CAS: 56-86-0 ✦ α-AMINOGLUTARIC ACID ✦ *l*-2-AMINOGLUTARIC ACID ✦ 2-AMINOPENTANEDIOIC ACID ✦ 1-AMINOPROPANE-1,3-DICARBOXYLIC ACID ✦ GLUSATE ✦ GLUTACID ✦ GLUTAMIC ACID ✦ α-GLUTAMIC ACID ✦ *d*-GLUTAMIENSUUR ✦ GLUTAMINIC ACID ✦ *l*-GLUTAMINIC ACID ✦ GLUTAMINOL ✦ GLUTATON

l-GLUTAMIC ACID HYDROCHLORIDE

Products and Uses: Used in beer as a flavor enhancer; in cosmetics as a preservative, as well as in hair permanent products. Also considered a dietary supplement, flavoring agent, nutrient, salt substitute, or antioxidant (slows down spoiling).

Precautions: A general purpose additive that FDA states is harmless when used for the intended purposes within limitations.

Synonyms: CAS: 138-15-8 ✦ α-AMINOGLUTARIC ACID HYDROCHLORIDE ✦ *l*-2-AMINOGLUTARIC ACID HYDROCHLORIDE ✦ 2-AMINOPENTANEDIOIC ACID HYDROCHLORIDE ✦ 1-AMINOPROPANE-1,3-DICARBOXYLIC ACID HYDROCHLORIDE ✦ GLUTAMIC ACID HYDROCHLORIDE ✦ α-GLUTAMIC ACID HYDROCHLORIDE ✦ GLUTAMINIC ACID HYDROCHLORIDE ✦ *l*-GLUTAMINIC ACID HYDROCHLORIDE

GLYCERIN

Products and Uses: Used in baked goods, candy, marshmallows, tobacco, jelly beans, ice cream toppings, sauces, cosmetics, blushes, skin creams, hair spray, mouth wash, and toothpaste. It is a food additive, humectant (keeps products moist), plasticizer (in coatings of cheeses and sausages), and solvent.

Precautions: Mildly toxic by swallowing large amounts. Effects on the human body by swallowing are headache, nausea, or vomiting. A skin and eye irritant. In the form of mist it is a breathing irritant. GRAS (generally recognized as safe) by FDA when used in limited amounts for intended purposes.

Synonyms: CAS: 56-81-5 ✦ GLYCERIN, ANHYDROUS ✦ GLYCERINE ✦ GLYCERIN, SYNTHETIC ✦ GLYCERITOL ✦ GLYCEROL ✦ GLYCYL ALCOHOL ✦ GROCOLENE ✦ MOON ✦ 1,2,3-PROPANETRIOL ✦ STAR ✦ SUPEROL ✦ SYNTHETIC GLYCERIN ✦ 90 TECHNICAL GLYCERINE ✦ TRIHYDROXYPROPANE ✦ 1,2,3-TRIHYDROXYPROPANE

GLYCEROL ESTER OF PARTIALLY DIMERIZED ROSIN

Products and Uses: It is in chewing gum, confectionery, candy eggs (shell), tablet-form food supplements, fruits, and vegetables as a color diluent and as a masticatory substance in chewing gum base.

Precautions: FDA approves use in moderate amounts to accomplish the intended effects.

Synonyms: NONE FOUND.

GLYCEROL-LACTO PALMITATE

Products and Uses: An emulsifier (it stabilizes mixture and keeps it uniform) in cakes, fats (rendered animal), and whipped toppings.

Precautions: FDA approves use in moderate amounts for intended purposes.

Synonyms: NONE FOUND.

GLYCEROL-LACTO STEARATE

Products and Uses: Used as an emulsifier (it stabilizes mixture and keeps it uniform) in cake mixes, chocolate coatings, fats (rendered animal), shortening, and whipped vegetable oil toppings.

Precautions: Harmless when used for intended purposes.

Synonyms: NONE FOUND.

GLYCEROL MONOOLEATE

Products and Uses: As a defoamer, dispersing agent, emulsifier (it stabilizes mixture and keeps it uniform) in coffee whiteners, packaging materials, and vegetable oil. It is also used as a plasticizer.

Precautions: FDA approves use at moderate levels to accomplish the desired results.

Synonyms: NONE FOUND.

GLYCERYL MONOOLEATE

Products and Uses: Frequently in baking mixes, beverages (nonalcoholic), beverage bases, chewing gum, and meat products as a flavor additive and agent.

Precautions: Harmless when used for intended purposes.

Synonyms: CAS: 25496-72-4

GLYCERYL TRISTEARATE

Products and Uses: It is an additive in chocolate (imitation), cocoa, confections, fats, and oil. Uses include as a crystallization accelerator, fermentation aid, formulation aid, fractionation aid, lubricant, release agent, surface-finishing agent, and winterization agent.

Precautions: FDA approves use within stated limits to accomplish the desired results.

Synonyms: CAS: 555-43-1

GLYCINE

Products and Uses: Used in beverage bases, beverages, fats (rendered animal), and cosmetic texturizers. It is a nutrient; it reduces the bitter taste of saccharin and prevents fats from getting rancid,

Precautions: Mildly toxic by swallowing. GRAS (generally recognized as safe) when used within FDA stated limits.

Synonyms: CAS: 56-40-6 ✦ AMINOACETIC ACID ✦ GLYCOLIXIR ✦ HAMPSHIRE GLYCINE

GLYCYRRHIZIN

Products and Uses: It is extracted from licorice root and is extremely sweet. Used as a tobacco humectant (prevents drying out), root beer foaming agent, in confectionery, chocolate, cocoa, and chewing gum. It masks the taste in pharmaceuticals such as aspirin. Used as a sweetener, nutrient, and flavoring.

Precautions: Harmless when used for intended purposes.

Synonyms: NONE FOUND.

GOLD

Products and Uses: Used in dental alloys, jewelry, industrial products, and medical applications.

Precautions: Harmless when used for intended purposes. Frequently suspected of causing allergic reaction when the actual cause is the alloy metal such as nickel.

Synonyms: CAS: 7440-57-5

GRAPHITE

Products and Uses: Used in bricks, shingles, lubricants, paints, coatings, pencil lead, makeup pencils, and cosmetic products as a polish, lubricant, filler, and as cement in industrial and consumer products.

Precautions: In powder form it is a fire risk.

Synonyms: CAS: 7782-42-5 ✦ BLACK LEAD ✦ PLUMBAGO

GUANIDINE CARBONATE

Products and Uses: A moisturizer in soap and cosmetic products.

Precautions: Toxic by swallowing.

Synonyms: NONE FOUND.

GUANINE

Products and Uses: Pearly coloration essence for cosmetics and toiletries in pearl finish nail polish, eye shadow, and lipstick. It has been replaced by synthetic chemicals such as bismuth.

Precautions: Harmless when used for intended purposes. It is derived from guano, sugar beets, yeast, clover seed, and fish scales.

Synonyms: CAS: 73-40-5 ✦ 2-AMINO-6-OXYPURINE ✦ 2-AMINOHYPOXANTHINE ✦ MEARLMAID

GUANYLIC ACID SODIUM SALT

Products and Uses: Used in canned foods, poultry, sauces, snack items, and soups as a flavor enhancer.

Precautions: Mildly toxic by swallowing large amounts.

Synonyms: CAS: 5550-12-9 ✦ DISODIUM GMP ✦ DISODIUM-5′-GMP ✦ DISODIUM GUANYLATE (FCC) ✦ DISODIUM-5′-GUANYLATE ✦ GMP DISODIUM SALT ✦ 5′-GMP DISODIUM SALT ✦ GMP SODIUM SALT ✦ SODIUM GMP ✦ SODIUM GUANOSINE-5′-MONOPHOSPHATE ✦ SODIUM GUANYLATE ✦ SODIUM-5′-GUANYLATE

GUAR GUM

Products and Uses: Commonly used in baked goods, baking mixes, beverages, cereals (breakfast), cheese, dairy product analogs, fats, gravies, ice cream, jams, jellies, milk products, oil, sauces, soup mixes, soups, sweet sauces, syrups, toppings, vegetable juices, vegetables (processed), and cheese spreads. Also used in cosmetic products and creams as an emulsifier, firming agent, formulation aid, stabilizer, and thickening agent.

Precautions: In large amounts it is mildly toxic by swallowing. GRAS (generally recognized as safe) when used within FDA limits.

Synonyms: CAS: 9000-30-0 ✦ A-20D ✦ BURTONITE V-7-E ✦ CYAMOPSIS GUM ✦ DEALCA TP1 ✦ DECORPA ✦ GALACTASOL ✦ GENDRIV 162 ✦ GUAR ✦ GUAR FLOUR ✦ GUM CYAMOPSIS ✦ GUM GUAR ✦ INDALCA AG ✦ JAGUAR No. 124 ✦ JAGUAR GUMA-20-D ✦ JAGUAR PLUS ✦ LYCOID DR ✦ REGONOL ✦ REIN GUARIN ✦ SUPERCOL U POWDER

GUM GHATTI

Products and Uses: An ingredient in beverage mixes, beverages, buttered syrup, and oil. It is an emulsifier (stabilizes and maintains mixes).

Precautions: Mildly toxic by swallowing large amounts. GRAS (generally recognized as safe) when used within FDA limits.

Synonyms: CAS: 9000-28-6 ✦ INDIAN GUM

GUM GUAIAC

Products and Uses: Used in fats (rendered animal) as an antioxidant (slows down spoiling) and preservative.

Precautions: Moderately toxic by swallowing gross amounts. Harmless when used within limits for intended purposes.

Synonyms: GUAIAC GUM

HALLUCINOGEN

Products and Uses: Found in natural plants, synthetic narcotics, and alkaloids for illegal, addictive, hallucinatory effects.

Precautions: Addictive drugs that act on the body and cause mental disturbance, imaginary experiences, coma, and possibly death. Sale and possession (except by physicians) is illegal in U.S.

Synonyms: CANNABIS ✦ MARIJUANA ✦ HASHISH ✦ LYSERGIC ACID (LSD) ✦ AMPHETAMINES ✦ MORPHINE DERIVATIVES

HELIUM

Products and Uses: Used for welding, balloon inflation (weather, research, party), diver's breathing equipment, and inflated advertising signs.

Precautions: A simple asphyxiant. Nonflammable gas. Discovered by Lockyer in 1868.

Synonyms: CAS: 7440-59-7 ✦ HELIUM, COMPRESSED (DOT) ✦ HELIUM, REFRIGERATED LIQUID (DOT)

HENNA

Products and Uses: Commonly found in hair dyeing products and also in antifungal medications. It is a red coloring agent.

Precautions: Always follow directions carefully when using products containing this item. Derived from the dried leaves of tropical plants.

Synonyms: NONE FOUND.

HEMATOXYLIN

Products and Uses: In inks and used as a colorant.

Precautions: A possible carcinogen (may cause cancer).

Synonyms: CAS: 517-28-2

HEPTACHLOR

Products and Uses: A pesticide for insects. Used as a termiticide (kills termites).

Precautions: Suspected carcinogen (may cause cancer). A poison by swallowing and skin contact. Acute exposure and chronic doses have caused liver damage. See also closely related chlordane. In humans, a dose of one to three grams can cause serious symptoms, especially where liver impairment is involved. Symptoms include tremors, convulsions, kidney damage, respiratory collapse, and death. EPA has canceled use of pesticides containing heptachlor except for exterior below grade application for termiticide.

Synonyms: CAS: 76-44-8 ✦ AGROCERES ✦ 3-CHLOROCHLORDENE ✦ DRINOX ✦ 3,4,5,6,7,8,8-HEPTACHLORODICYCLOPENTADIENE ✦ HEPTAGRAN ✦ HEPTAMUL ✦ RHODIACHLOR

n-HEPTADECANOL

Products and Uses: Various applications include perfumes, shaving products, shaving lotions, colognes, soaps, toiletries, and cosmetics. It is used in the manufacture of detergents; also as a plasticizer, fixative, and wetting agent.

Precautions: Harmless when used for intended purposes.

Synonyms: NONE FOUND.

HEPTANAL

Products and Uses: An ingredient in perfumes, fragrances, foods, beverages, ice creams, candy, and bakery products. It affects the taste or smell of final product. Used as a synthetic fruit and nut flavoring.

Precautions: Mildly toxic by swallowing of large amounts.

Synonyms: CAS: 111-71-7 ✦ ENANTHAL ✦ ENANTHALDEHYDE ✦ ENANTHOLE ✦ HEPTALDEHYDE ✦ OENANTHALDEHYDE ✦ OENANTHOL

n-HEPTANOIC ACID

Products and Uses: Used in lubricants and brake fluids for slipperiness and reducing friction.

Precautions: Mildly toxic by swallowing.

Synonyms: CAS: 111-14-8 ✦ ENANTHIC ACID ✦ n-HEPTYLIC ACID ✦ HEPTOIC ACID

HEPTYL ACETATE

Products and Uses: Used as a fruit essence flavoring.

Precautions: Irritating to skin, eyes, nose, and throat. Harmless when used for intended purposes.

Synonyms: CAS: 112-06-1 ✦ HEPTANYL ACETATE ✦ ACETATE C-7 ✦ 1-HEPTYL ACETATE

HEPTYL ALCOHOL

Products and Uses: Used in cosmetics and perfumery for formulating and as a solvent.

Precautions: Moderately toxic by swallowing and skin contact.

Synonyms: CAS: 111-70-6 ✦ 1-HEPTANOL ✦ ENANTHYL ALCOHOL

HEPTYL FORMATE

Products and Uses: Utilized in baked goods, beverages, candy, and ice cream as an artificial fruit-essence flavoring.

Precautions: FDA approves use at moderate levels to accomplish the intended effect. A skin irritant.

Synonyms: CAS: 112-23-2 ✦ FORMIC ACID, HEPTYL ESTER ✦ HEPTANOL, FORMATE ✦ n-HEPTYL METHANOATE

HEPTYL HEPTOATE

Products and Uses: An ingredient in food and beverage products. Used as an artificial fruit-essence flavoring.

Precautions: Harmless when used for intended purposes.

Synonyms: NONE FOUND.

HEPTYL PARABEN

Products and Uses: An additive in beer, beverages (fermented malt), fruit drinks (noncarbonated), wine, and soft drinks (noncarbonated). Used as an antioxidant (maintains freshness) or preservative.

Precautions: Harmless when used for intended purposes within designated limits.

Synonyms: n-HEPTYL p-HYDROXYBENZOATE ✦ HEPTYL ESTER OF PARA-HYDROXYBENZOIC ACID

HEPTYL PELARGONATE

Products and Uses: Frequently used in foods, beverages, colognes, and perfumes for flavoring and fragrancing.

Precautions: Harmless when used for intended purposes.

Synonyms: NONE FOUND.

HESPERIDIN

Products and Uses: Used in assorted food and beverage items as a synthetic sweetener.

Precautions: Extracted from citrus fruit peel. Can cause allergic reaction.

Synonyms: CAS: 520-26-3 ✦ CIRANTIN ✦ VITAMIN P

HEXACHLOROBENZENE

Products and Uses: A wood preservative used on outdoor furniture, fences, foundations, and landscape timbers.

Precautions: A carcinogen (causes cancer). Toxic by swallowing.

Synonyms: CAS: 118-74-1 ✦ PERCHLOROBENZENE

HEXACHLOROPHENE

Products and Uses: Limited use in germicidal soaps and foot powders as a bactericide; once utilized in many cleansing products.

Precautions: FDA prohibits use unless prescribed by a physician.

Synonyms: CAS: 70-30-4

HEXAHYDROBENZOIC ACID

Products and Uses: Used in paints, varnishes, dry-cleaning soaps, oils, and detergents as a drier, solvent, lubricant, and stabilizer.

Precautions: Label directions should be followed very carefully for products that contain this chemical.

Synonyms: CYCLOHEXANECARBOXYLIC ACID ✦ NAPTHENIC ACID

HEXAMETHYLENETRAMINE

Products and Uses: An additive in adhesives (rubber to textile), fungicide, textiles (shrink proofing), and antibacterial.

Precautions: A skin irritant.

Synonyms: CAS: 100-97-0 ✦ METHENAMINE ✦ HMTA ✦ AMINOFORM ✦ HEXAMINE

1-HEXANAL

Products and Uses: A fruit and coconut flavoring in beverages, desserts, candies, bakery products, and gels; used in bactericides and perfumes for flavoring and fragrancing.

Precautions: Mildly toxic by swallowing large amounts and by breathing fumes. An irritant to skin and eyes. FDA approves use at moderate levels to accomplish the intended effect.

Synonyms: CAS: 66-25-1 ✦ ALDEHYDE C-6 ✦ CAPROALDEHYDE ✦ CAPROIC ALDEHYDE ✦ CAPRONALDEHYDE ✦ n-CAPROYLALDEHYDE ✦

1,2,6-HEXANETRIOL

Products and Uses: An ingredient in skin lotions and cosmetic makeup as a softener and moisturizer.

Precautions: Mildly toxic by swallowing. An eye and skin irritant.

Synonyms: CAS: 106-69-4 ✦ HEXANE-1,2,6 TRIOL ✦ HEXANETRIOL-1,2,6

HEXAPHOS

Products and Uses: A water softener and soap deposit preventative in cleansers, laundry soaps, dishwashing powders, and detergents.

Precautions: Follow label directions carefully on product containing this chemical.

Synonyms: GLASSY PHOSPHATE

HEXAZINONE

Products and Uses: A herbicide or weed killer.

Precautions: Moderately toxic by swallowing. Mildly toxic by skin contact. An eye irritant. USDA rates it as a minimal hazard.

Synonyms: CAS: 51235-04-2 ✦ 3-CYCLOHEXYL-6-(DIMETHYLAMINO)-1-METHYL-s-TRIAZINE-2,4(1H,3H)-DIONE ✦ 3-CYCLOHEXYL-6-(DIMETHYLAMINO)-1-METHYL-1,3,5-TRIAZINE-2,4(1H,3H)-DIONE ✦ DPX 3674 ✦ VELPAR ✦ VELPAR WEED KILLER

HEXENOL

Products and Uses: A fragrancing or scenting additive that occurs naturally in grasses, leaves, herbs, and teas. Also used in perfumes and odorants.

Precautions: Could cause allergic reaction. Harmless when used for intended purposes.

Synonyms: 3-HEXEN-1-OL ✦ LEAF ALCOHOL

HEXETIDINE

Products and Uses: An ingredient in bactericides, fungicides, and algicides that is also used for synthetic textiles.

Precautions: Harmless when directions are followed and product is used for intended purposes.

Synonyms: CAS: 141-94-6 ✦ (AMINO-1,3-BIS[β-ETHYLHEXYL]-5-METHYLHEXAHYDROPYRIMIDINE

HEXYL ACETATE

Products and Uses: A synthetic fruit and berry taste enhancer used for beverages, desserts, candies, bakery products, and flavored gum.

Precautions: In large amounts it is mildly toxic by swallowing. FDA approves use at moderate levels to accomplish the desired results.

Synonyms: CAS: 142-92-7 ✦ ACETIC ACID HEXYL ESTER ✦ n-HEXYL ACETATE ✦ 1-HEXYL ACETATE ✦ HEXYL ALCOHOL, ACETATE ✦ HEXYL ETHANOATE

HEXYL ALCOHOL

Products and Uses: Used in pharmaceuticals, antiseptics, perfumes, toiletries, and textiles. It is a plasticizer, solvent, intermediate, and finishing agent.

Precautions: Moderately toxic by swallowing and skin contact. A skin and severe eye irritant.

Synonyms: CAS: 111-27-3 ✦ 1-HEXANOL ✦ AMYL CARBINOL

HEXYL CINNAMALDEHYDE

Products and Uses: A jasmine-type odor for fruit, berry, beverage honey flavorings, desserts, candies, and bakery products; used as a flavoring, aroma, and taste enhancer.

Precautions: Moderately toxic by swallowing gross amounts. A skin irritant. Harmless when used for intended effects.

Synonyms: CAS: 101-86-0 ✦ α-HEXYLCINNAMALDEHYDE ✦ HEXYL CINNAMIC ALDEHYDE ✦ α-HEXYLCINNAMIC ALDEHYDE ✦ α-n-HEXYL-β-PHENYLACROLEIN ✦ 2-(PHENYLMETHYLENE)OCTANOL

HEXYLENE

Products and Uses: An additive used in flavors, perfumes, and dyes.

Precautions: Moderately toxic irritant to skin, eyes, nose, and throat. It could cause an allergic reaction.

Synonyms: CAS: 592-41-6 ✦ 1-HEXENE

HEXYLENE GLYCOL

Products and Uses: Various applications include brake fluids, inks, cosmetics, fabric finishing or sizing, fuel and lubricant additive; prevents ice forming in carburetors. Used as an additive and emulsifier (maintains mixes in suspension).

Precautions: Moderately toxic by swallowing. Mildly toxic by skin contact. Effects on the human body by breathing are conjunctiva (eye), olfactory (nasal), and pulmonary (lung) changes.

Synonyms: CAS: 107-41-5 ✦ 4-METHYL-2,4-PENTANEDIOL

HEXYLRESORCINOL

Products and Uses: An antibacterial and anthelmintic (destroys and expels worms) in medications and pharmaceuticals.

Precautions: Irritant to respiratory tract, eye, and skin. Concentrated solutions can cause burns of the skin, nose, and throat.

Synonyms: CAS: 136-77-6 ✦ 1,3-DIHYDROXY-4-HEXYLBENZENE

HIGH-FRUCTOSE CORN SYRUP

Products and Uses: Typically found in beverages (carbonated), candy, confections, desserts (frozen), drinks (dairy), fruits (canned), ham (chopped or processed), hamburger, ice cream, luncheon meat, meat loaf, poultry, and sausage as a flavoring or sweetening (nutritive) agent.

Precautions: GRAS (generally recognized as safe). Harmless when used for intended purposes.

Synonyms: CORN SYRUP, HIGH-FRUCTOSE

HOMOCYSTEINE

Products and Uses: A naturally occurring amino acid that is present in excessive amounts in people who drink more than nine cups of coffee a day.

Precautions: High levels are believed to damage arteries and trigger atherosclerosis (hardening of the arteries). Folic acid, a B vitamin, rids the body of homocysteine. The recommended dose of folic acid is 400 mcgs per day.

Synonyms: NONE KNOWN.

HOMOMENTHYL SALICYLATE

Products and Uses: Used in suntan lotions, creams, oils, toiletries, protectants, and cosmetics to filter or absorb UV (ultraviolet) radiation of sunlight.

Precautions: Could cause allergic reaction. Harmless when used for intended purposes.

Synonyms: 3,3,5-TRIMETHYLCYCLOHEXYL SALICYLATE

HOMOSALATE

Products and Uses: A UV (ultraviolet) filter or screen in suntan lotions, creams, oils, protectants, and cosmetics.

Precautions: Could cause allergic reaction. Harmless when used for intended purposes.

Synonyms: NONE FOUND.

HOP EXTRACT, MODIFIED

Products and Uses: As a flavoring agent for beverages, for beer, fruit, and root beer.

Precautions: Harmless when used for intended purposes.

Synonyms: MODIFIED HOP EXTRACT.

HOPS OIL

Products and Uses: An aromatic, spicy, bitter beverage flavoring. In whiskey, beverages, desserts, bakery products, and seasonings.

Precautions: GRAS (generally recognized as safe) when used at a moderate level to accomplish the intended effects.

Synonyms: NONE FOUND.

HUMIC ACID

Products and Uses: Used in printing ink pigments, cosmetic facial mud baths, and as a coloring, softening, moisturizing agent.

Precautions: Harmless when used for intended purposes. It is a natural stream pollutant and is thought to be capable of triggering the red tide phenomenon due to microorganisms in seawater.

Synonyms: NONE FOUND.

HYDRARGAPHEN

Products and Uses: A germicide, bactericide, and fungicide for wool, adhesives, fabrics, leather, paints, and wood products.

Precautions: A poison. A severe eye irritant.

Synonyms: CAS: 14235-86-0 ✦ HYDRAPHEN ✦ CONOTRANE ✦ PENOTRANE ✦ SEPTOTAN

HYDRATED ALUMINUM OXIDES

Products and Uses: In paper, polishes, adhesives, inks, paints, and cosmetics as a coloring agent.

Precautions: Harmless when used for intended purposes.

Synonyms: NONE FOUND.

HYDRAZINE SULFATE

Products and Uses: It is a fungicide and germicide. Known to destroy fungi and bacteria.

Precautions: A carcinogen (causes cancer). A poison by swallowing. Effects on the human body by swallowing are paresthesia (abnormal sensations), somnolence (sleepiness), nausea, or vomiting. A mutagen (changes inherited characteristics). An eye irritant.

Synonyms: CAS: 10034-93-2 ✦ DIAMINE SULFATE ✦ DIAMIDOGEN SULFATE

HYDRIODIC ACID

Products and Uses: An expectorant (aids in loosening secretions) in cough syrups. Used as a disinfectant and as a germicide.

Precautions: In large amounts it is poison by swallowing and by breathing fumes. A corrosive and poisonous irritant to skin, eyes, nose, and throat.

Synonyms: CAS: 10034-85-2 ✦ HYDROGEN IODIDE

HYDROCHLORIC ACID

Products and Uses: Used in hair bleaching products, swimming pool chemicals, toilet cleaner, drain cleaner, as a buffer, a neutralizing agent, and an etching agent. It is an oxidizer in hair rinses and color remover in hair products. Also used in food processing.

Precautions: A human poison. Mildly toxic to humans by breathing fumes. A corrosive irritant to the skin, eyes, nose, and throats. A concentration of 35 ppm causes irritation of the throat after short exposure. On EPA Extremely Hazardous Substance list.

Synonyms: CAS: 7647-01-0 ✦ CHLOROHYDRIC ACID ✦ HYDROCHLORIC ACID, ANHYDROUS ✦ HYDROCHLORIC ACID, SOLUTION INHIBITED ✦ HCL ✦ HYDROGEN CHLORIDE ✦ HYDROGEN CHLORIDE, ANHYDROUS ✦ HYDROGEN CHLORIDE, REFRIGERATED LIQUID ✦ MURIATIC ACID ✦ SPIRITS OF SALT

HYDROCINNAMIC ACID

Products and Uses: An additive in perfumes and fragrances. It is used as a flavoring and scenting agent.

Precautions: Moderately toxic by swallowing. Mildly toxic by skin contact. A skin irritant. FDA approves use at a moderate levels to accomplish the intended effect.

Synonyms: CAS: 122-97-4 ✦ 3-BENZENEPROPANOL ✦ HYDROCINNAMYL ALCOHOL ✦ (3-HYDROXYPROPYL)BENZENE ✦ γ-PHENYLPROPANOL ✦ 3-PHENYLPROPANOL ✦ PHENYLPROPYL ALCOHOL ✦ γ-PHENYLPROPYL ALCOHOL ✦ 3-PHENYLPROPYL ALCOHOL

HYDROCOLLOID

Products and Uses: Various applications include in food and cosmetic products for emulsifiers, thickeners, and gelling agents. Imparts smoothness and texture to products.

Precautions: Harmless when used for intended purposes.

Synonyms: GUM ARABIC ✦ AGAR ✦ GUAR GUM ✦ STARCHES ✦ DEXTRAN ✦ GELATIN

HYDROGENATED VEGETABLE OIL

Products and Uses: Used in margarine, shortening, and many processed foods as a flavoring, seasoning, cooking, and frying oil.

Precautions: Diets high in fats and oils cause heart disease, obesity, and possibly cancer.

Synonyms: NONE FOUND.

HYDROGEN PEROXIDE

Products and Uses: Used as a bleaching product for skin and hair, also in cosmetic creams, mouthwashes, dentifrices, and hair wave products. It is an antiseptic, bleaching, oxidizing agent, and preservative. In the food industry it has applications in cheese whey (annatto-colored), corn syrup, distilling materials, eggs (dried), emulsifiers containing fatty acid esters, herring, milk, packaging materials, starch, tea (instant), tripe, whey, wine, and wine vinegar.

Precautions: Moderately toxic by breathing fumes, swallowing large amounts, and by skin contact. A corrosive irritant to skin, eyes, nose, and throat. A human mutagen (changes inherited characteristics). Questionable carcinogen (causes cancer). FDA approves use within limitations.

Synonyms: CAS: 7722-84-1 ✦ ALBONE ✦ DIHYDROGEN DIOXIDE ✦ HIOXYL ✦ HYDROGEN DIOXIDE ✦ HYDROGEN PEROXIDE, SOLUTION (over 52% peroxide) ✦ HYDROGEN PEROXIDE, STABILIZED (over 60% peroxide) ✦ HYDROPEROXIDE ✦ INHIBINE ✦ OXYDOL ✦ PERHYDROL ✦ PERONE ✦ PEROXAN ✦ PEROXIDE ✦ SUPEROXOL

HYDROQUINONE

Products and Uses: Various applications include uses in cosmetic skin bleaching creams, paints, varnishes, fuels, oils, and photographic developers.

Precautions: A human poison by swallowing. An active allergen and a strong skin irritant. The swallowing of one gram may cause nausea, dizziness, a sense of suffocation, increased respiration, vomiting, pallor, muscle twitching, headache, delirium, and collapse.

Synonyms: CAS: 123-31-9 ✦ QUINOL ✦ HYDROQUINOL ✦ p-DIHYDROXYBENZENE

HYDROQUINONE DIMETHYL ETHER

Products and Uses: A component in paints, perfumes, dyes, cosmetics, and suntan products. Also used as a weathering agent, fixative, and flavoring (pleasant, sweet, clover odor).

Precautions: Poison by swallowing. Human reproductive effects (infertility or sterility or birth defects).

Synonyms: CAS: 654-42-2 ✦ DMB ✦ 1,4-DIMETHOXYBENZENE ✦ DIMETHYL HYDROQUINONE

HYDROXYACETIC ACID

Products and Uses: Used in fabric dyes, leather dyes, adhesives, cleaning compound, and polishing compound.

Precautions: Harmless when used for intended purposes.

Synonyms: GLYCOLIC ACID

m-HYDROXYBENZALDEHYDE

Products and Uses: For beverages, ices, desserts, candies, bakery products, spices and liqueurs. In disinfectants, fumigants, and perfumery. Used as a buttery, nutlike seasoning. Possesses anti-inflammatory properties.

Precautions: Moderately toxic by swallowing and skin contact. A skin irritant. Could cause allergic reaction.

Synonyms: CAS: 90-2-8 ✦ SALICYLALDEHYDE ✦ SALICYLIC ALDEHYDE ✦ 2-FORMYLPHENOL ✦ SALICYLAL

p-HYDROXYBENZOIC ACID ETHYL ESTER

Products and Uses: An antimicrobial agent and preservative in baked goods, beverages, food colors, and wine.

Precautions: In large amounts it is moderately toxic by swallowing. GRAS (generally recognized as safe) when used within limitations stated by FDA.

Synonyms: CAS: 120-47-8 ✦ ASEPTOFORM E ✦ BONOMOLD OE ✦ p-CARBETHOXYPHENOL ✦ EASEPTOL ✦ ETHYL-p-HYDROXYBENZOATE ✦ ETHYL PARABEN ✦ ETHYL PARASEPT ✦ p-HYDROXYBENZOIC ETHYL ESTER

2-HYDROXY ETHYL CARBAMATE

Products and Uses: A sizing and finishing agent for wash-and-wear cotton fabrics.

Precautions: Can cause allergic reactions. Generally harmless when used for intended purposes.

Synonyms: CAS: 589-41-3 ✦ ETHYL-N-HYDROXYCARBAMATE ✦ HYDROXYCARBAMIC ACID ETHYL ESTER ✦ N-HYDROXYURETHANE

HYDROXY ETHYL CELLULOSE

Products and Uses: Used in papers and textiles as a thickening and suspending agent for products.

Precautions: Can cause allergic reactions. Generally harmless when used for intended purposes.

Synonyms: CELLOSIZE

HYDROXYLATED LECITHIN

Products and Uses: It is a clouding agent and emulsifier in bakery products, beet sugar, beverages (dry mix), margarine, and yeast.

Precautions: FDA approves use at moderate levels to accomplish the desired results.

Synonyms: NONE FOUND.

2-HYDROXY-3-METHYL-2-CYCLOPENTEN-1-ONE

Products and Uses: Used as a flavoring agent, food additive, and fragrancing agent in bakery products, beverages (nonalcoholic), chewing gum, confections, gelatin desserts, ice cream, puddings, and syrups.

Precautions: In large amounts it is moderately toxic by swallowing. A human mutagen (changes inherited characteristics).

Synonyms: CAS: 80-71-7 ✦ CORYLON ✦ CORYLONE ✦ CYCLOTEN ✦ MAPLE LACTONE ✦ 3-METHYLCYCLOPENTANE-1,2-DIONE ✦ METHYL CYCLOPENTENOLONE

p-HYDROXYPROPYL BENZOATE

Products and Uses: A food additive, preservative, and antimicrobial in baked goods, beverages, food colors, milk, sausage (dry), and wine.

Precautions: Mildly toxic by swallowing large amounts. An allergen. FDA approves use within stated limitations.

Synonyms: CAS: 94-13-3 ✦ ASEPTOFORM P ✦ BETACIDE P ✦ BONOMOLD OP ✦ 4-HYDROXYBENZOIC ACID PROPYL ESTER ✦ p-HYDROXYBENZOIC ACID PROPYL ESTER ✦ NIPASOL ✦ p-OXYBENZOESAUREPROPYLESTER ✦ PARABEN ✦ PARASEPT ✦ PASEPTOL ✦ PRESERVAL P ✦ PROPYL p-HYDROXYBENZOATE ✦ n-PROPYL p-HYDROXYBENZOATE ✦ PROPYLPARABEN ✦ PROPYLPARASEPT ✦ PROTABEN P ✦ TEGOSEPT P

HYDROXYPROPYL CELLULOSE

Products and Uses: An emulsifier, film former, stabilizer, suspending agent, and thickening agent. Used as a food additive, preservative, and antimicrobial.

Precautions: Slightly toxic by swallowing gross amounts. In glazes, oil, vitamin tablets, vitamin wafers, and whipped toppings.

Synonyms: CAS: 9004-64-2 ✦ HYDROXYPROPYL ETHER OF CELLULOSE ✦ KLUCEL

HYDROXYTRIPHENYLSTANNANE

Products and Uses: Its purpose is to kill fungus on food products. Used for beef, carrots, goat, horse, lamb, peanut hulls, peanuts, pecans, pork, potatoes, and sugar beet roots.

Precautions: In large amounts it is a poison by swallowing. A severe eye irritant.

Synonyms: CAS: 76-87-9 ✦ DU-TER ✦ FENOLOVO ✦ FENTIN HYDROXIDE ✦ HAITIN ✦ HYDROXYTRIPHENYLTIN ✦ SUZU H ✦ TPTH ✦ TRIPHENYLTIN OXIDE ✦ TUBOTIN

HYPOCHLOROUS ACID

Products and Uses: Utilized in bleaching, purifying, cleaning, and germ killing; also for textile bleaches, water purification, pool cleaners, and antiseptics.

Precautions: An irritant to skin and eyes.

Synonyms: CAS: 7790-92-3 ✦ HYPOCHLORITE SOLUTION ✦ CALCIUM HYPOCHLORITE ✦ SODIUM HYPOCHLORITE

ICELAND MOSS

Products and Uses: Added to sea biscuits to prevent weevil infestation. Used in hair-setting lotions, personal care products, and cosmetic bases as a filler or additive, and for convalescent food. It is also an alcoholic beverage flavoring and a suspension agent for hair products and toiletries.

Precautions: Harmless when used for intended purposes.

Synonyms: SCANDINAVIAN MOSS ✦ ICELAND LICHEN

ICHTHAMMOL

Products and Uses: An additive in pharmaceutical products such as skin ointments, preparations, cosmetics, toiletries, and special dermatological soaps. Considered an antiseptic, emollient (skin softener), and demulcent (relieves skin irritation).

Precautions: Harmless when used for intended purposes.

Synonyms: AMMONIUM ICHTHOSULFONATE ✦ BITUMOL ✦ BITUMINOL ✦ ICHTHIUM ✦ ICHTOPUR ✦ LITHOL ✦ PETROSULPHOL ✦ TUMENOL

INDOLE

Products and Uses: A compound found in broccolli and other vegetables. Indications are that it protects lab animals from developing breast and stomach cancer. This ingredient is also used in perfumes for consumer toiletries and personal care products. (Unpleasant odor in high concentrations but diluted solutions are pleasant).

Precautions: Moderately toxic by swallowing and skin contact. FDA approves use at moderate levels to accomplish the intended effect.

Synonyms: CAS: 120-72-9 ✦ 1-AZAINDENE ✦ 1-BENZAZOLE ✦
BENZOPYRROLE ✦ 2,3-BENZOPYRROLE ✦ KETOLE

INOSITOL

Products and Uses: Derived from vegetables, citrus fruits, cereal grains, liver, kidney, and other meats. In dietary supplement of vitamin B and in medications.

Precautions: GRAS (generally recognized as safe). Harmless when used for intended purposes.

Synonyms: CAS: 87-89-8 ✦ cis-1,2,3,5-trans-4,6-CYCLOHEXANEHEXOL ✦
i-INOSITOL ✦ MESO-INOSITOL

INVERT SOAPS

Products and Uses: Used in detergents and soaps for disinfecting and cleansing.

Precautions: Harmless when used for intended purposes.

Synonyms: NONE FOUND.

INVERT SUGAR

Products and Uses: Usually in candy, icings, soft drinks, brewing, and as humectant (moisturizer that prevents drying). It is a 50–50 mixture of two sugars, dextrose and fructose. It is sweeter and more soluble than table sugar. Primarily a sweetener (nutritive) and moisturizing additive.

Precautions: GRAS (generally recognized as safe). Found in low-quality foods, causes tooth decay and should be avoided.

Synonyms: CAS: 8013-17-0 ✦ INVERT SUGAR SYRUP

IODINE

Products and Uses: An additive in dyes, antiseptics, germicides, pharmaceuticals, medicinal soap, a germicide in skin cosmetics, and OTC (over-the-counter) sore throat medicine and cough syrups. Derived from various sources, from mines to kelp. Found in hundreds of products. As internal medication for iodine deficiencies, external medication for its disinfecting and antibacterial properties. Interestingly, consuming excessive amounts of raw cabbage can cause an iodine deficiency and affect thyroid production in people with existing low-iodine intakes.

Precautions: Toxic by swallowing large amounts and by breathing fumes. Strong irritant to eyes and skin. Could cause allergic reaction.

Synonyms: CAS: 7553-56-2 ✦ JOD ✦ IODE ✦ IODINE CRYSTALS

IODINE TINCTURE

Products and Uses: In solution of iodine and potassium iodide in alcohol. An antiseptic. Used as a germicide.

Precautions: Toxic by swallowing. Should not be used on open cuts because of toxicity.

Synonyms: NONE FOUND.

IPECAC

Products and Uses: A medical emetic (causes vomiting) and expectorant used to induce vomiting after swallowing toxic products.

Precautions: Toxic. Label directions must be followed carefully.

Synonyms: CEPHALIS IPECACUANHA

IRON

Products and Uses: In enriched baked goods, cereal products, flour, pasta, and milk it is used as a dietary or nutritional supplement.

Precautions: Very beneficial in appropriate amounts and uses. In gross amounts it is a possible carcinogen (may cause cancer) that caused tumor growth in animals. Iron is potentially toxic in all forms and by all routes of exposure. Iron supplements are the leading cause of accidental poisoning deaths for children under six, who often mistake the pills for candy. Chronic exposure to excess levels of iron can result in pathological deposits of iron in the body tissues, the symptoms of which are fibrosis of the pancreas, diabetes mellitus, and liver cirrhosis.

Synonyms: CAS: 7439-89-6 ✦ ANCOR EN 80/150 ✦ ARMCO IRON ✦ CARBONYL IRON ✦ IRON, CARBONYL ✦ IRON, ELECTROLYTIC ✦ IRON, ELEMENTAL ✦ IRON, REDUCED

IRON BLUE

Products and Uses: Found in paints, printing inks, artists' colors, cosmetic eye shadow, laundry blueing, dyes, and finishes. Used as a pigment or coloring agent.

Precautions: Harmless when used for intended purposes.

Synonyms: NONE FOUND.

IRON OXIDE

Products and Uses: In pigments, cat food, dog food, and packaging materials as a color additive, cosmetic color, and constituent of paperboard.

Precautions: FDA approves use at limited levels to accomplish the desired results.

Synonyms: CAS: 1309-37-1 ✦ BAUXITE RESIDUE ✦ BLACK OXIDE OF IRON ✦ BLENDED RED OXIDES OF IRON ✦ BURNTISLAND RED ✦ BURNT SIENNA ✦ BURNT UMBER ✦ CALCOTONE RED ✦ COLCOTHAR ✦ COLLOIDAL FERRIC OXIDE ✦ FERRIC OXIDE ✦ INDIAN RED ✦ IRON(III) OXIDE ✦ IRON OXIDE RED ✦ IRON SESQUIOXIDE ✦ JEWELER'S ROUGE ✦ MARS BROWN ✦ MARS RED ✦ NATURAL IRON OXIDES ✦ NATURAL RED OXIDE ✦ OCHRE ✦ PRUSSIAN BROWN ✦ RED IRON OXIDE ✦ RED OCHRE ✦ ROUGE ✦ RUBIGO ✦ SIENNA ✦ SYNTHETIC IRON OXIDE ✦ VENETIAN RED ✦ VITRIOL RED ✦ YELLOW OXIDE OF IRON

IRON OXIDE RED

Products and Uses: A coloring material in marine paints, metal primers, polishing compounds, theatrical rouge, blush, cosmetics, and grease paints.

Precautions: Harmless when used for intended purposes.

Synonyms: CAS: 1332-37-2 ✦ BURNT SIENNA ✦ INDIAN RED ✦ RED IRON OXIDE ✦ RED OXIDE ✦ ROUGE ✦ TURKEY RED

ISOAMYL FORMATE

Products and Uses: In candy, dessert gels, ice cream, and puddings it is an artificial fruit-essence flavoring.

Precautions: In large amounts it is moderately toxic by swallowing. A skin irritant. Concentrated material is very irritating and can cause narcosis (numbing). The symptoms are usually brief, but it is possible upon severe or prolonged exposure to cause serious consequences. FDA approves use at moderate levels to accomplish the desired results.

Synonyms: CAS: 110-45-2 ✦ FORMIC ACID, ISOPENTYL ESTER ✦ ISOAMYL METHANOATE ✦ ISOPENTYL ALCOHOL, FORMATE ✦ ISOPENTYL FORMATE ✦ 3-METHYLBUTYL FORMATE

ISOBORNYL ACETATE

Products and Uses: A popular pine needle odorant and essence in toilet waters, bubble bath, bath oils, antiseptics, air fresheners, soaps, and food flavorings.

Precautions: Could cause allergic reaction.

Synonyms: NONE FOUND.

ISOBORNYL SALICYLATE

Products and Uses: A fixative and filter in preparations for perfumes, toiletries, cosmetics, suntan preparations, oils, lotions, mousses, and gels.

Precautions: A possible allergen in susceptible individuals.

Synonyms: NONE FOUND.

ISOBUTYL ACETATE

Products and Uses: Used as a sealant; topcoat for lacquers. In foods it is a fruit-like flavoring agent.

Precautions: In excessive amounts it is mildly toxic by swallowing and by breathing fumes. A skin and eye irritant. Approved by the FDA for use at moderate levels to accomplish the desired results.

Synonyms: CAS: 110-19-0 ✦ ACETIC ACID, ISOBUTYL ESTER ✦ ACETIC ACID-2-METHYLPROPYL ESTER ✦ 2-METHYLPROPYL ACETATE ✦ 2-METHYL-1-PROPYL ACETATE ✦ β-METHYLPROPYL ETHANOATE

ISOBUTYL-p-AMINOBENZOATE

Products and Uses: Additive in topical anesthetic and suntan oils, creams, lotions, gels, mousses and preparations.

Precautions: Harmless when used for intended purposes.

Synonyms: CAS: 94-14-4 ✦ p-AMINOBENZOIC ACID ISOBUTYL ESTER ✦ (2-METHYL)-p-AMINOBENZOATE

ISOBUTYLENE-ISOPRENE COPOLYMER

Products and Uses: Utilized in chewing gum, gumballs, sealants, coatings, finishings, cements, and adhesives

Precautions: Harmless when used for intended purposes. FDA approves use in amounts to accomplish the intended effect.

Synonyms: BUTYL RUBBER

ISOBUTYL STEARATE

Products and Uses: Used as a waterproofer, in coatings, polishes, inks, cosmetics, makeup cream, blush, and soaps.

Precautions: Harmless when used for intended purposes.

Synonyms: CAS: 646-13-9

ISOBUTYRIC ACID

Products and Uses: Used as a disinfecting agent, solvent, and seasoning. It is in varnish, flavorings, and perfume bases.

Precautions: A poison by swallowing. Moderately toxic by skin contact. A corrosive irritant to the eyes, skin, nose, and throat.

Synonyms: CAS: 79-31-2 ✦ DIMETHYLACETIC ACID ✦ ISOPROPYLFORMIC ACID ✦ α-METHYLPROPIONIC ACID ✦ 2-METHYLPROPANOIC ACID ✦ 2-METHYLPROPIONIC ACID

ISODECYL CHLORIDE

Products and Uses: Multiple uses include applications in oils, fats, greases, pharmaceuticals, and detergents, cleaning compounds and solvents.

Precautions: Harmless when used for intended purposes.

Synonyms: NONE FOUND.

ISOEUGENOL

Products and Uses: Usually included in room deodorants, toiletries, colognes, hair tonics, and hand creams. Fragrance variously described as spice-clove, floral-carnation, or vanillin.

Precautions: Moderately toxic by swallowing. A human mutagen (changes inherited characteristics).

Synonyms: CAS: 97-54-1 ✦ 1-HYDROXY-2-METHOXY-4-PROPENYLBENZENE ✦ 4-HYDROXY-3-METHOXY-1-PROPENYLBENZENE ✦ 2-METHOXY-4-PROPENYLPHENOL ✦ 4-PROPENYLGUAIACOL

ISOPROPANOLAMINE

Products and Uses: An emulsifying agent (stabilizes and maintains mixes to aid in suspension of oily liquid) in dry-cleaning soaps, wax removers, and cosmetics.

Precautions: Harmless when used for intended purposes.

Synonyms: CAS: 78-96-6 ✦ 2-HYDROXYPROPYLAMINE ✦ 1-AMINO-2-PROPANOL ✦ MIPA

ISOPROPYL ACETATE

Products and Uses: A solvent and aromatic in paints, lacquers, inks, and perfumes.

Precautions: Moderately toxic by swallowing. Mildly toxic by breathing. Harmful effects on the human body by breathing vapors. Narcotic in high concentration. Chronic exposure can cause liver damage.

Synonyms: CAS: 108-21-4 ✦ ACETIC ACID ISOPROPYL ESTER ✦ 2-ACETOXYPROPANE ✦ 2-PROPYL ACETATE

ISOPROPYL ALCOHOL

Products and Uses: Component in lotions, antifreeze, shellac solvent, quick-drying inks, toiletries, body lotions, after-shave lotions, hair preparations, and cosmetics. Used as a deicer, preservative, solvent, and dehydrating (drying) agent.

Precautions: Can cause corneal burns and irritation. Moderately toxic to humans. Mildly toxic by skin contact. Effects by swallowing or breathing: flushing, pulse rate decrease, blood pressure lowering, anesthesia, narcosis, headache, dizziness, mental depression, hallucinations, distorted perceptions, dyspnea (shortness of breath), respiratory depression, nausea, vomiting, and coma. A mutagen (changes inherited characteristics).

Synonyms: CAS: 67-63-0 ✦ DIMETHYLCARBINOL ✦ ISOHOL ✦ ISOPROPANOL ✦ LUTOSOL ✦ PETROHOL ✦ PROPAN-2-OL ✦ 2-PROPANOL ✦ sec-PROPYL ALCOHOL ✦ i-PROPYLALKOHOL ✦ SPECTRAR

ISOPROPYLAMINE

Products and Uses: Utilized as a solvent, insecticide, germicide, among others in pharmaceuticals, dyes, and bactericides.

Precautions: Poison by skin contact in concentrated form. Moderately toxic by swallowing. Mildly toxic by breathing. A severe skin and eye irritant. A narcotic in high concentrations. Could cause allergic response.

Synonyms: CAS: 75-31-0 ✦ 2-AMINOPROPANE ✦ MONOISPROPYLAMINE ✦ 2-PROPANAMINE

ISOPROPYL CITRATE

Products and Uses: An additive in margarine and vegetable oils. A preservative and sequestrant (affects the final product's appearance, flavor, or texture).

Precautions: FDA states GRAS (generally recognized as safe) when used within limitations.

Synonyms: NONE FOUND.

ISOPROPYL ETHER

Products and Uses: Used in paint remover, varnish remover, spot remover, and rubber cements.

Precautions: Mildly toxic by swallowing, breathing, and skin contact. A skin irritant.

Synonyms: CAS: 108-20-3 ✦ DIISOPROPYL ETHER ✦ DIISOPROPYL OXIDE ✦ 2-ISOPROPOXYPROPANE

ISOPROPYL MYRISTATE

Products and Uses: Its purpose is to preserve (it prevents rancidity). It is not easily soluble in water and maintains long-lasting effects. It also improves absorption through skin in cosmetic creams, toiletries, suntan lotions, and topical skin medications.

Precautions: Could be a skin irritant and a possible comodgenic (clogs pores). Derived from coconut oil.

Synonyms: CAS: 110-27-0 ✦ ISOMYST ✦ KESSCOMIR ✦ TETRADECANOIC ACID, ISOPROPYL

ISOPROPYL PALMITATE

Products and Uses: A common ingredient additive in cosmetic lotions, creams, deodorants, and skin care products. Useful as an emollient (skin softener), and emulsifer (stabilizes and suspends oils in mixture).

Precautions: Could cause allergic reaction.

Synonyms: CAS: 142-91-6 ✦ CRODAMOL IPP ✦ DELTYL ✦ HEXANDECANOIC ACID ✦ ISOPROPYL ESTER ✦ ISOPAL ✦ PROPAL ✦ TEGESTER ISOPALM ✦ ISOPROPYL HEXADECANOATE ✦ DELTYL PRIME ✦ STEPAN D-70 ✦ UNIMATE IPP ✦ WICKENOL 111

JAPAN WAX

Products and Uses: A substitute for beeswax, used for candles, floor waxes, polishes, back plasters, ointments, furniture polish, and dental impression compounds.

Precautions: Harmless when used for intended purposes.

Synonyms: JAPAN TALLOW ✦ SUMAC WAX

JASMINE OIL

Products and Uses: An odorant in perfume in colognes, toiletries, bath products, talcs, and hair products.

Precautions: Could cause allergic reaction.

Synonyms: JASMINE FLOWER ✦ SWEET JASMINE OIL

JEWELER'S ROUGE

Products and Uses: A red powdered haematite, iron oxide. Because it is a mild abrasive it is used in metal cleaners and polishes.

Precautions: Not for internal use.

Synonyms: HAEMATITE IRON OXIDE

JOJOBA OIL

Products and Uses: A substitute for sperm oil used for transmission lubricants, carnauba wax and beeswax, cosmetic preparations, hair products, shampoos, conditioners, moisturizers, skin products, toiletries, softeners, hand lotions, and suntan products.

Precautions: Possible allergic reactions in susceptible individuals.

Synonyms: DESERT OIL ✦ OIL OF JOJOBA

JUNIPER BERRY OIL

Products and Uses: A seasoning and flavoring for assorted food and beverage products such as gin, root beer, wintergreen, ginger, and liqueurs.

Precautions: Mildly toxic by swallowing large amounts. A skin irritant. An allergen. Effects on the body by swallowing are severe kidney irritation similar to that caused by turpentine. GRAS (generally recognized as safe) when used at moderate levels to accomplish the intended effect.

Synonyms: CAS: 8012-91-7 ✦ OIL OF JUNIPER BERRY ✦ WACHOLDERBEER OEL (GERMAN)

KAOLIN

Products and Uses: A mineral named for the location in China where first found. Many uses include cements, fertilizers, cosmetic face powder, baby powder, bath powder, foundation facial makeup, paints, and antidiarrheal medications. Useful in wine filtration, and as anticaking agent, clarifying agent, covering agent, and filling agent.

Precautions: Harmless when used for intended purposes.

Synonyms: CAS: 1332-58-7 ✦ ALTOWHITES ✦ BENTONE ✦ CONTINENTAL ✦ DIXIE ✦ EMATHLITE ✦ FITROL ✦ FITROL DESICCITE 25 ✦ GLOMAX ✦ HYDRITE ✦ KAOPAOUS ✦ KAOPHILLS-2 ✦ LANGFORD ✦ MCNAMEE ✦ PARCLAY ✦ PEERLESS ✦ SNOW TEX

KARAYA GUM

Products and Uses: Used in baked goods, candy (soft), frozen dairy desserts, milk products, and toppings. It is an emulsifier (maintains mixes to aid in suspension of oily liquids), stabilizer (maintains consistency), thickening agent, and adhesive.

Precautions: In excessive amounts it is very mildly toxic by swallowing. A mild allergen. FDA approves when used within limitations.

Synonyms: CAS: 9000-36-6 ✦ STERCULIA GUM ✦ INDIA TRAGACANTH ✦ KADAYA GUM

KELP

Products and Uses: A large coarse seaweed found off the coast of California. An algae used in chewing gum base, fertilizer, and dietary supplements as a filler, seasoning, or flavoring.

Precautions: Harmless when used for intended purposes. However, some people who consumed large amounts experienced enlarged thyroids. FDA states it is GRAS (generally recognized as safe).

Synonyms: *MACROCYSTIS PYRIFERAE*

KERATINASE

Products and Uses: Found in lotion and cream cosmetic products. Useful as a depilatory (hair remover).

Precautions: A potential skin irritant and allergen. Directions must be followed carefully on products containing this enzyme.

Synonyms: NONE FOUND.

KEROSENE

Products and Uses: A fuel for lanterns, lamps, heaters, flares, and stoves. Used as a degreaser, solvent, cosmetic ingredient, combustible, and in insect sprays.

Precautions: A possible carcinogen. A severe skin irritant. A mutagen (changes inherited characteristics). Effects on the body by swallowing are somnolence (sleepiness), hallucinations and distorted perceptions, coughing, nausea, vomiting, and fever. Aspiration of vomitus has caused chemical pneumonia especially in children. Moderately explosive in the form of vapor when exposed to heat or flame.

Synonyms: CAS: 8008-20-6 ✦ COAL OIL ✦ DEOBASE ✦ STRAIGHT-RUN KEROSENE

KETONE

Products and Uses: Used in lacquers and paints as a solvent and for textile processing.

Precautions: Moderately toxic by various routes. A skin and severe eye irritant. Harmful effects on the body by breathing. Container directions must be carefully followed.

Synonyms: NONE FOUND.

KOJIC ACID

Products and Uses: Ingredient in insecticides, antifungals, and antimicrobial agents.

Precautions: Moderately toxic by swallowing.

Synonyms: CAS: 501-30-4 ✦ [5-HYDROXY-2-(HYDROXYMETHYL)-4-PYRONE]

LACQUER

Products and Uses: Cellulose esters or ethers in coating materials used on metal, protective, or decorative coatings; also in paper products, textiles, plastics, furniture polish, and nail polish.

Precautions: Variable toxicity depending upon concentrations of products. May cause allergic reactions. A dangerous fire hazard due to the highly flammable solvents commonly used. A severe explosion hazard in the form of vapor when exposed to flame.

Synonyms: NITROCELLULOSE-ALKYD LACQUERS ✦ NITROCELLULOSE ✦ VINYL RESINS ✦ ACRYLIC RESINS

LACTATED MONO-DIGLYCERIDES

Products and Uses: An emulsifier (stabilizes mixes to aid in suspension of oily liquids) and stabilizer (maintains uniform consistency) in margarine.

Precautions: Harmless when used for intended purposes.

Synonyms: NONE FOUND.

LACTIC ACID

Products and Uses: Produced from milk by the action of lactic acid bacteria when milk is fermented to make cheese. It is also found in other sour foods such as sauerkraut, fermented meats, molasses and is a preservative with some pickled foods such as pearl onions and olives. It is also a normal component of our body. It is produced be muscle activity and normal metabolism and is in our blood and urine. It is a component of all plant and animal tissues and it is impossible to keep out of our diet.Useful as an acid, antimicrobial agent, curing agent, flavor enhancer, flavoring agent, pH control agent, pickling agent, solvent, and as a vehicle for other chemicals.

Precautions: Moderately toxic by swallowing of gross quantities. A severe skin and eye irritant. GRAS (generally recognized as safe) when used at moderate levels to accomplish the desired results.

Synonyms: CAS: 50-21-5 ✦ ACETONIC ACID ✦ ETHYLIDENELACTIC ACID ✦ 1-HYDROXYETHANECARBOXYLIC ACID ✦ 2-HYDROXYPROPANOIC ACID ✦ 2-HYDROXYPROPIONIC ACID ✦ α-HYDROXYPROPIONIC ACID ✦ KYSELINA MLECNA ✦ *dl*-LACTIC ACID ✦ MILCHSAURE ✦ MILK ACID ✦ ORDINARY LACTIC ACID ✦ RACEMIC LACTIC ACID

LACTOSE

Products and Uses: Found in infant foods, pharmaceutical bases, laxatives, baked goods, candies, margarine, and butter. It is also used in the manufacture of penicillin.

Precautions: Could cause allergic reaction. Harmless when used for intended purposes.

Synonyms: CAS: 63-42-3 ✦ LACTIN ✦ LACTOBIOSE ✦ *d*-LACTOSE ✦ MILK SUGAR ✦ SACCHARUM LACTIN

LACTYLIC ESTERS of FATTY ACIDS

Products and Uses: Used in baked goods, bakery mixes, frozen desserts, fruit juices (dehydrated), fruits (dehydrated), milk or cream substitutes for beverage coffee, pancake mixes, pudding mixes, rice (precooked instant), shortening (liquid), vegetable juices (dehydrated), and vegetables (dehydrated). It is an emulsifier (stabilizes and maintains mixes to aid in suspension of oily liquids).

Precautions: Harmless when used for intended purposes. FDA approves use at a moderate level to accomplish the desired results.

Synonyms: NONE FOUND.

LAMPBLACK

Products and Uses: A pigment for cements, inks, crayons, polishes, carbon paper, soap, matches, and cosmetic eyeliner pencils. For coloring, coating, lubricating, and polishing.

Precautions: Harmless when used for intended purposes while taking normal precautions.

Synonyms: CARBONACEOUS MATERIAL ✦ MICROCRYSTALLINE CARBON

LANOLIN

Products and Uses: The oil produced by hair glands in sheep. Used in ointments, leather polishes, face creams, facial tissue, hairdressing products, cosmetics, lipstick, mascara, rouge, eyeshadow, and suntan preparations. Lanolin anhydrous is used in chewing gum as a base. Primarily, it is a moisturizer.

Precautions: A frequent cause of allergic reactions to the skin.

Synonyms: WOOL FAT

LARD

Products and Uses: Used in baked foods, fried foods, pharmacy ointments, pomades, soaps, shaving creams, and cosmetics for moisturizing.

Precautions: High in saturated fats. Otherwise harmless when used for intended purposes.

Synonyms: HOG FAT ✦ STEARIN ✦ PALMITIN ✦ OLEIN ✦ PORK FAT ✦ PORK OIL

LATEX

Products and Uses: The milky fluid derived from sap of rubber trees, also produced synthetically. Used in gloves, adhesives, medical equipment, undergarments, and personal products for flexibility, strengthening, and waterproofing of consumer items.

Precautions: Could cause allergic reaction to skin that comes in contact with latex. Sixteen deaths attributed to latex in 1990. Simple allergy results in skin welts, reddening and blistering as well as other respiratory problems. Severe problems with fatal reactions can occur from breathing problems and blood pressure drop. It can lead to anaphylaxis. Some believe the allergy is from the chemicals added in processing.

Synonyms: RUBBER HYDROCARBON ✦ STYRENE-BUTADIENE COPOLYMER ✦ ACRYLATE RESINS ✦ POLYVINYL ACETATE ✦ NATURAL LATEX

LAURIC ACID

Products and Uses: An additive in coconut oil, laurel oil, vegetable fats, soaps, detergents, and flavorings in foods. Used as a lubricant, soap, foam, or bubble agent.

Precautions: In large amounts it is mildly toxic by swallowing.

Synonyms: CAS: 143-07-7 ✦ DODECANOIC ACID ✦ DODECOIC ACID ✦ DUODECYLIC ACID ✦ HYDROFOL ACID ✦ HYSTRENE ✦ LAUROSTEARIC ACID ✦ NEO-FAT ✦ NINOL EXTRA ✦ 1-UNDECANECARBOXYLIC ACID ✦ WECOLINE

LAUROYL PEROXIDE

Products and Uses: An ingredient in fats, oils, waxes, cleaners, polishes, for furniture and flooring. Used as a bleaching or drying agent.

Precautions: Toxic by swallowing and breathing. Corrosive to eyes, nose, and throat.

Synonyms: CAS: 105-74-8 ✦ DODECANOLY PEROXIDE ✦ ALPEROX C ✦ LAUROX ✦ LAURYDOL

LAURYL SULFOACETATE

Products and Uses: A common additive in toothpaste, tooth powder, liquid dentifrices, foaming bath products, shampoos, and synthetic detergents. Used as a biodegradable detergent, wetting, scouring, emulsifying, dispersing, and foaming agent.

Precautions: Harmless when used for intended purposes.

Synonyms: LANTHANOL ✦ SODIUM LAURYL SULFOACETATE

LAVENDER OIL

Products and Uses: Various applications include bakery products, beverages (non-alcoholic), chewing gum, confections, ice cream products, in toiletries, shaving products, mouthwashes, breath fresheners, toothpaste, toothpowder, and cologne. It is a seasoning or fragrancing agent.

Precautions: In excessive amounts it is mildly toxic by swallowing. A skin irritant. GRAS (generally recognized as safe) when used at moderate levels to accomplish the desired results.

Synonyms: CAS: 8000-28-0 ✦ LAVENDEL OEL (GERMAN) ✦ OIL OF LAVENDER

LEAD

Products and Uses: A heavy metal used in paint pigment, solder, pool cue chalk, crayons from China, glazes on ceramic dishes and bowls, among other products. Prohibited from interstate commerce since the middle 1970s. It is still manufactured and used locally. Lead has been found in wines, possibly from the foil used on bottles. (Pencil "lead" is not lead at all; rather it is a mixture of graphite and clay.) Prevents bottom growth on boat hulls and rust development on metal. Used as a filler and for radiation protection.

Precautions: Suspected carcinogen. Poison by swallowing. Effects on the human body by swallowing and breathing are loss of appetite, anemia, malaise, insomnia, headache, irritability, muscle pains, joint pains, tremors, hallucinations, distorted perceptions, muscle weakness, gastritis, and liver changes. The major organ systems affected are the nervous system, blood system, and kidneys.

Any amount of lead is unsafe for children, pregnant women, and nursing mothers. Studies now suggest that blood levels of lead below 10 mg/dL can have the effect of lowering the IQ scores of children. Low levels of lead impair neurotransmission and immune system function and may increase systolic blood pressure. Kidney damage can occur from exposure.

Severe toxicity can cause sterility, abortion, neonatal mortality, and morbidity (diseased state). A mutagen (changes inherited characteristics). Very heavy intoxication can sometimes be detected by formation of a dark line on the gum margins, the so-called "lead line." For the general population, exposure to lead occurs from breathed air, dust of various types, and food and water with an approximate 50–50 division between breathing and swallowing routes. Lead occurs in water in either dissolved or particulate form. Commonly found in paint chips from older properties where children live. Children eat paint chips and develop chronic symptoms. Acidic foods (tomatoes, fruit juices, and so on) leach lead out of decorative glazes in chinaware and cause chronic symptoms in those who ingest food or drinks.

Synonyms: CAS: 7439-92-1 ✦ GLOVER ✦ LEAD FLAKE ✦ LEAD S2 ✦ OLOW (POLISH) ✦ OMAHA ✦ OMAHA & GRANT ✦ SI ✦ SO

LECITHIN

Products and Uses: Used in baked goods, beverage powders, cocoa powder, fat (griddling), fillings, chocolate, meat products, margarine, poultry, shortening, cosmetic products, lotions, soaps, creams, eye makeup, and face makeup. It is an antioxidant (slows reaction with oxygen, slows spoiling), emulsifier (stabilizes and maintains mixes to aid in suspension of oily liquids).

Precautions: GRAS (generally recognized as safe) when used within FDA limitations.

Synonyms: NONE FOUND.

LEMONGRASS OIL

Products and Uses: A flavoring and odorant ingredient in bakery products, beverages (nonalcoholic), chewing gum, confections, gelatin desserts, ice cream, puddings, soaps, and toiletry articles.

Precautions: Mildly toxic by swallowing excessive amounts. A skin irritant. GRAS (generally recognized as safe) by FDA when used at moderate levels to accomplish the desired results.

Synonyms: CAS: 8007-02-1 ✦ GUATEMALA LEMONGRASS OIL ✦ MADAGASCAR LEMONGRASS OIL ✦ WEST INDIAN LEMONGRASS OIL

LEVULOSE

Products and Uses: Utilized in baked goods and beverages (low calorie) as a formulation aid, sweetener (nutritive), and processing oil.

Precautions: Harmless when used for intended purposes.

Synonyms: CAS: 7660-25-5 ✦ FRUCTOSE (FCC) ✦ FRUIT SUGAR ✦ FRUTABS ✦ LAEVORAL ✦ LAEVOSAN ✦ LEVUGEN

LICORICE ROOT EXTRACT

Products and Uses: A flavor enhancer in bacon, baked goods, beverages (alcoholic), beverages (nonalcoholic), candy (hard), candy (soft), chewing gum, cocktail mixes, herbs, ice cream, plant protein products, seasonings, soft drinks, syrups, vitamin and mineral dietary supplements, and whipped products (imitation).

Precautions: In large amounts it is mildly toxic by swallowing. Excessive consumption linked to water retention and high blood pressure and loss of potassium, which in turn leads to irregular heartbeat and possible paralysis. Possible pyelonephritis (kidney disease) from excessive ingestion also. GRAS (generally recognized as safe) when used within FDA limitations.

Synonyms: CAS: 8008-94-4 ✦ GLYCYRRHIZA ✦ GLYCYRRHIZAE (LATIN) ✦ GLYCYRRHIZA EXTRACT ✦ GLYCYRRHIZINA ✦ KANZO (JAPANESE) ✦ LICORICE ✦ LICORICE EXTRACT ✦ LICORICE ROOT

LIME, SULFURATED

Products and Uses: Used in medications, depilatories, luminous paints, and inks.

Precautions: An irritant and caustic.

Synonyms: CALCIUM SULFIDE, CRUDE

LINALOOL

Products and Uses: A fruity or floral-scented fragrance in toilet waters, shaving lotions, hand lotions, and colognes.

Precautions: Could cause allergic reactions.

Synonyms: CAS: 78-70-6 ✦ LINALOL ✦ EX BOIS DE ROSE OIL, SYNTHETIC ✦ BERGAMOL OIL ✦ FRENCH LAVENDER

LINOLEIC ACID

Products and Uses: Used in infant formula and margarine as a dietary supplement or flavoring agent.

Precautions: A human skin irritant. Swallowing can cause nausea and vomiting. GRAS (generally recognized as safe).

Synonyms: CAS: 60-33-3 ✦ LEINOLEIC ACID ✦ 9,12-LINOLEIC ACID ✦ cis,cis-9,12-OCTADECADIENOIC ACID ✦ cis-9,cis-12-OCTADECADIENOIC ACID ✦ 9,12-OCTADECADIENOIC ACID

LINOLEYTRIMETHYLAMMONIUM BROMIDE

Products and Uses: An ingredient in germicides, deodorants, algicides, and slime control products.

Precautions: A severe skin and eye irritant. Highly toxic by swallowing. Label directions should be carefully followed.

Synonyms: NONE FOUND.

LINSEED OIL

Products and Uses: A drying agent, thickener, and suspension for various products in paints, adhesives, varnishes, putty, printing inks, soaps, and pharmaceutical items. Useful as a drying agent, thickener, and suspension for various products.

Precautions: A common allergen and skin irritant.

Synonyms: CAS: 8001-26-1 ✦ GROCO ✦ L-310 ✦ FLAXSEED OIL

LIPOXIDASE

Products and Uses: Used as a bleaching or whitening agent in bread products: rolls, buns, breads, and bakery goods.

Precautions: Harmless when used for intended purposes.

Synonyms: NONE FOUND.

LITHIUM STEARATE

Products and Uses: An emulsifier (aids in suspension of oily liquids), lubricant, and plasticizer in cosmetics, skin creams, facial makeups, plastics, waxes, and greases, varnishes, lacquers, and lubricants.

Precautions: Harmless when used for intended purposes.

Synonyms: NONE FOUND.

LOCUST BEAN GUM

Products and Uses: An emulsifier (aids in suspension of oily liquids), stabilizer (keeps consistency uniform), and thickening agent in baked goods, beverage bases (nonalcoholic), beverages, candy, cheese, fillings, gelatins, ice cream, jams, jellies, pies, puddings, and soups. Also used in cosmetics, textile sizing, textile finishing, and pharmaceuticals.

Precautions: Mildly toxic by swallowing. GRAS (generally recognized as safe) within stated FDA limits.

Synonyms: CAS: 9000-40-2 ✦ ALGAROBA ✦ CAROB BEAN GUM ✦ CAROB FLOUR ✦ NCI-C50419 ✦ ST. JOHN'S BREAD ✦ SUPERCOL

LSD

Products and Uses: Infinitesimal amounts on paper put on tongue can cause deadly detrimental effects. An illegal, habit-forming hallucinogen. Use often causes unintentional suicide of user.

Precautions: An illegal controlled substance that is habit forming. It is a strong hallucinogen that elicits optical (visual) or auditory (hearing) hallucinations, depersonalization (person feels mind and body are separated), perceptual disturbances (cannot judge distances), and disturbances of thought processes (cannot control body at will).

Synonyms: CAS: 50-37-3 ✦ *D*-LYSERGIC ACID DIETHYLAMIDE

LYCOPENE

Products and Uses: The main pigment (coloring matter) in tomatoes, paprika, and rose hips. It is a carotenoid. Research indicates it is a possible cancer preventitive.

Precautions: Harmless when used for intended purposes.

Synonyms: CAS: 502-65-8 ✦ CAROTENE

MACASSER OIL

Products and Uses: In hair tonic, oils, and conditioning treatments.

Precautions: Derived from Indian nut kernels. Harmless when used for intended purposes. Could cause allergic reaction in susceptible individuals.

Synonyms: KUSUM OIL ✦ KON OIL ✦ PAKA OIL ✦ CEYLON OAK OIL

MACE

Products and Uses: A spray and aerosol tear gas used as a riot control agent and in self-defense products.

Precautions: Strong irritant to eyes and tissue as gas or liquid.

Synonyms: CAS: 532-27-4 ✦ MACE ✦ CHEMICAL MACE ✦ PHENACYCLCHLORIDE ✦ PHENYL CHLOROMETHYL KETONE β-CHLOROACETOPHENONE

MAGNESITE

Products and Uses: A white, colorless, or grey mineral found in mixes (dry) and table salt. It is an alkali, anticaking agent, carrier, color-retention agent, and drying agent.

Precautions: Harmless when used for intended purposes.

Synonyms: CAS: 546-93-0 ✦ CARBONATE MAGNESIUM ✦ CARBONIC ACID, MAGNESIUM SALT ✦ DCI LIGHT MAGNESIUM CARBONATE ✦ HYDROMAGNESITE ✦ MAGMASTER ✦ MAGNESIA ALBA ✦ MAGNESIUM CARBONATE ✦ MAGNESIUM(II) CARBONATE ✦ MAGNESIUM CARBONATE, PRECIPITATED ✦ STAN-MAG MAGNESIUM CARBONATE

MAGNESIUM

Products and Uses: An essential dietary mineral for women. Can ease the symptoms of PMS (premenstrual syndrome) and improve energy in people with CFS (chronic fatigue syndrome). Recent studies have suggested magnesium routinely given to pregnant women to treat high blood pressure and premature labor may also sharply reduce the risk of cerebral palsy and retardation. Intravenously it can prevent heart attack in patients with heart disease. Occurs naturally in green leaves, nuts, cereals, grains, and seafood.

Precautions: Harmless when used for intended purposes in appropriate amounts.

Synonyms: CAS: 7439-95-4 ✦ Mg ✦ ATOMIC NUMBER 12

MAGNESIUM AMMONIUM PHOSPHATE

Products and Uses: Used as a fire retardant for fabrics, canvas products, tents, camping equipment, and firefighter apparel.

Precautions: Harmless when used for intended purposes.

Synonyms: MAGNESIUM AMMONIUM ORTHOPHOSPHATE

MAGNESIUM CARBONATE

Products and Uses: An additive in inks, dentifrices, cosmetics, free-running table salt, antacid, and foods. Used as a drying, color-retaining, and anticaking agent. It is the synthetic form of mineral magnesite.

Precautions: Harmless when used for intended purposes.

Synonyms: MAGNESITE (NATURALLY OCCURRING MATERIAL)

MAGNESIUM CHLORIDE

Products and Uses: Multiple uses include disinfectants, fire extinguishers, fireproofing wood, and floor sweeping compounds. Used for meat (raw cuts), and poultry (raw cuts). It is a color-retention agent, firming agent, flavoring agent, and tissue softening agent (tenderizer) in foods.

Precautions: Moderately toxic by swallowing. A human mutagen (changes inherited characteristics). In humid environments it causes steel to rust very rapidly. GRAS (generally recognized as safe) when used within the FDA limitations.

Synonyms: CAS: 7786-30-3 ✦ DUS-TOP

MAGNESIUM FLUOSILICATE

Products and Uses: Used for waterproofing, mothproofing, and in laundry products, textiles, canvasses, denims, awnings, and fabrics.

Precautions: Poison by swallowing.

Synonyms: CAS: 18972-56-0 ✦ MAGNESIUM SILICOFLUORIDE ✦ MAGNESIUM HEXAFLUOROSILICATE

MAGNESIUM LAURYL SULFATE

Products and Uses: Common chemical in detergents, cleaning products, and cosmetics. Used as a foaming, wetting, and emulsifying (aids in suspension of oily liquids) agent.

Precautions: Harmless when used for intended purposes.

Synonyms: NONE FOUND.

MAGNESIUM PHOSPHATE, MONOBASIC

Products and Uses: Used for fireproofing and in wood treatment products.

Precautions: Harmless when used for intended purposes.

Synonyms: CAS: 13092-66-5 ✦ MAGNESIUM BIOPHOSPHATE ✦ ACID MAGNESIUM PHOSPHATE ✦ MAGNESIUM TETRAHYDROGEN PHOSPHATE

MAGNESIUM PHOSPHATE, TRIBASIC

Products and Uses: An antacid, food additive dietary supplement, a dentifrice and cleaning agent.

Precautions: Harmless when used for intended purposes.

Synonyms: CAS: 7757-86-0 ✦ MAGNESIUM PHOSPHATE, NEUTRAL ✦ TRIMAGNESIUM PHOSPHATE

MAGNESIUM RICINOLEATE

Products and Uses: Used in cosmetic products.

Precautions: Could cause allergic reaction in susceptible individuals. Harmless when used for intended purposes.

Synonyms: NONE FOUND.

MAGNESIUM SILICATE HYDRATE

Products and Uses: In table salt it is an anticaking agent and filter aid.

Precautions: A human skin irritant. GRAS (generally recognized as safe) when used within FDA limitations.

Synonyms: CAS: 1343-90-4

MAGNESIUM STEARATE

Products and Uses: Various uses include in candy, gum, mints, and sugarless gum. Also used in dusting powder, medicines, and as cosmetic emulsifier (maintains mixes and aids in suspension of oils). It is an anticaking agent, binder, emulsifier (aids in suspension of oily liquids), lubricant, nutrient, processing aid, release agent, and stabilizer.

Precautions: Harmless when used for intended purposes. GRAS (generally recognized as safe). Must conform to FDA specifications and within limited amounts.

Synonyms: CAS: 557-04-0

MALATHION

Products and Uses: A pesticide in animal feed, cattle feed concentrate blocks (nonmedicated), citrus pulp (dehydrated), grapes, packaging materials, and safflower oil. Used as an insecticide for flies, insects, head lice, and mosquitos.

Precautions: A human poison by swallowing and skin contact. Can penetrate intact skin. Effects on humans by swallowing are coma, blood pressure depression, and difficulty in breathing. A possible carcinogen. A human mutagen (changes inherited characteristics). Has caused allergic sensitization of the skin. FDA approves use within limitations.

Synonyms: CAS: 121-75-5 ✦ CALMATHION ✦ CARBETHOXY MALATHION ✦ CARBETOVUR ✦ CARBETOX ✦ CARBOFOS ✦ CARBOPHOS ✦ CELTHIGN ✦ CHEMATHION ✦ CIMEXAN ✦ CYTHION ✦ EMMATOS ✦ EMMATOS EXTRA ✦ S-ESTER with O,O-DIMETHYL PHOSPHOROTHIOATE ✦ ETHIOLACAR ✦ ETIOL ✦ EXTERMATHION ✦ FORMAL ✦ FORTHION ✦ FOSFOTHION ✦ FOSFOTION ✦ FYFANON ✦ HILTHION ✦ KARBOFOS ✦ KOP-THION ✦ KYPFOS ✦ MALACIDE ✦ MALAFOR ✦ MALAGRAN ✦ MALAKILL ✦ MALAMAR ✦ MALAPHELE ✦ MALAPHOS ✦ MALASOL ✦ MALASPRAY ✦ MALATOX ✦ MALDISON ✦ MALMED ✦ MALPHOS ✦ MALTOX ✦ MALTOX MLT ✦ MERCAPTOSUCCINIC ACID DIETHYL ESTER ✦

MERCAPTOTHION ✦ MLT ✦ MOSCARDA ✦ OLEOPHOSPHOTHION ✦ ORTHO MALATHION ✦ PHOSPHORODITHIOIC ACID-O,O-DIMETHYL ESTER-S-ESTER with DIETHYL MERCAPTOSUCCINATE ✦ PHOSPHOTHION ✦ PRIODERM ✦ SADOFOS ✦ SADOPHOS ✦ SUMITOX ✦ TAK ✦ VEGFRU MALATOX ✦ VETIOL ✦ ZITHIOL

MALIC ACID

Products and Uses: Commonly used in beverages (dry mix), candy (hard), candy (soft), chewing gum, fats (chicken), fillings, fruit juices, fruits (processed), gelatins, jams, jellies, lard, nonalcoholic beverages, puddings, shortening, soft drinks, and wine. It is an acidifier, flavor enhancer, and flavoring agent.

Precautions: Moderately toxic by swallowing large amounts. A skin and severe eye irritant. GRAS (generally recognized as safe) when used within FDA limitations. Could cause allergic reaction.

Synonyms: CAS: 6915-15-7 ✦ HYDROXYSUCCINIC ACID ✦ KYSELINA JABLECNA

MANGANESE GLUCONATE

Products and Uses: Used as food additive, dietary supplement, and nutrient. In baked goods, beverages (nonalcoholic), dairy product analogs, fish products, meat products, milk products, vitamin tablets, and poultry products.

Precautions: GRAS (generally recognized as safe).

Synonyms: CAS: 6485-39-8

MANNITOL

Products and Uses: A sweetener and dietetic food base used as a thickener, stabilizer, flavoring agent, and lubricant.

Precautions: Mildly toxic by swallowing excessive amounts. A human mutagen (changes inherited characteristics).

Synonyms: CAS: 69-65-8 ✦ MANNA SUGAR ✦ MANNITE ✦ OSMITROL ✦ 1,2,3,4,5,6-HEXANEHEXOL

MARIJUANA

Products and Uses: Used in smoking materials. It is a hallucinogenic illegal street drug. It does have proven medical applications for controlling nausea.

Precautions: A hallucinogen. It contains 400 compounds, some of which are carcinogenic. Moderately toxic by swallowing. An animal teratogen (causes abnormal fetus development), causes reproductive effects (infertility, or sterility, or birth defects). Human mutagen (changes inherited characteristics). Harmful effects include heart rate changes, blood pressure drop. When smoked or swallowed can cause delirium, drowsiness, weakness, and reflex weakness. Overdose can cause coma and death. This herb does have medical applications and can be very beneficial for epileptics and controlling nausea of chemotherapy patients.

Synonyms: CAS: 8063-14-7 ✦ CANNABIS ✦ MARY JANE ✦ DOPE ✦ INDIAN HEMP ✦ HASHISH

MARJORAM OIL

Products and Uses: Used in perfume and toilet waters as a flavoring agent or odorant. The Spanish grade is used in fish, meat, sauces, and soups.

Precautions: GRAS (generally recognized as safe) when used at moderate levels to accomplish the intended effect.

Synonyms: NONE FOUND.

MEAT TENDERIZER

Products and Uses: Bromelain, a pineapple plant enzyme used as meat tenderizer.

Precautions: Harmless when used as directed.

Synonyms: ANANASE ✦ EXTRANASE ✦ INFLAMEN ✦ PLANT PROTEASE CONCENTRATE

MENTHOL

Products and Uses: Utilized in cigarettes, perfumes, shaving creams, lotions, after-shave, hair products, foods, liqueurs, chewing gum, cold medications, and cough drops as a flavoring agent and odorant.

Precautions: In large amounts it is moderately toxic by swallowing. A severe eye irritant. FDA approves use at moderate levels to accomplish the intended effect.

Synonyms: CAS: 89-78-1 ✦ HEXAHYDROTHYMOL ✦ 2-ISOPROPYL-5-METHYL-CYCLOHEXANOL ✦ p-MENTHAN-3-OL ✦

1-MENTHOL ✦ 5-METHYL-2-(1-METHYLETHYL)CYCLOHEXANOL ✦
PEPPERMINT CAMPHOR

MERCERIZED COTTON

Products and Uses: A strengthening process by which threads, yarns, and cottons are passed through a solution of sodium hydroxide and washed with water while under tension. It causes fibers to shrink, resulting in strengthening and improving color properties.

Precautions: Harmless when used for intended purposes.

Synonyms: NONE FOUND.

MERCURIC ARSENATE

Products and Uses: Additive in paints for waterproofing and marine antifouling (prevents underwater growth) paint.

Precautions: A poison. A confirmed carcinogen (causes cancer). Label instructions for use must be strictly followed.

Synonyms: CAS: 7784-37-4 ✦ MERCURY ARSENATE ✦ MERCURY ORTHOARSENATE

MERCURIC CHLORIDE

Products and Uses: Used in fungicides, insecticides, wood preservatives, embalming fluid, and printing.

Precautions: Toxic by swallowing, breathing, and skin absorption. A poison.

Synonyms: CAS: 7487-94-7

MERCURIC CYANIDE

Products and Uses: Used in antiseptics and germicidal soaps for cleansing and antibacterial products.

Precautions: Toxic by swallowing, breathing, and skin absorption. Effects on the human body by swallowing are nausea, vomiting, diarrhea, and somnolence (sleepiness).

Synonyms: CAS: 592-04-1 ✦ MERCURY CYANIDE

MESCALINE

Products and Uses: Derived from a Mexican cactus and used in medical and bio-chemical research.

Precautions: A hallucinogenic drug. Moderately toxic by swallowing. Poison by intravenous route. Effects on the body by intramuscular route are hallucinations and distorted perceptions.

Synonyms: CAS: 54-04-6 ✦ MEZCALINE ✦ MEZCLINE ✦ 3,4,5-TRIMETHOXYBENZENEETHANIMINE ✦ 3,4,5-TRIMETHOXYPHENETHYLAMINE

MESITYL OXIDE

Products and Uses: A solvent, remover and repellant used in lacquers, inks, stains, paint, varnish remover, and insecticides.

Precautions: Moderately toxic by swallowing. Mildly toxic by breathing and skin contact. Irritating to the eyes. High concentrations are narcotic. Readily absorbed through skin. Prolonged exposure can injure liver, kidney, and lungs.

Synonyms: CAS: 141-79-7 ✦ METHYL ISOBUTENYL KETONE ✦ ISOPROPYLIDENEACETONE ✦ 4-METHYL-3-PENTEN-2-ONE

METHANE

Products and Uses: A natural gas and coal gas from decaying vegetation and other organic matter in swamps and marshes. In the form of natural gas, methane is used as a fuel. It is purchased by power companies from dump sites for home heating and cooking.

Precautions: Severe fire and explosion hazard; forms explosive mixture with air (5 to 15% by volume).

Synonyms: CAS: 74-82-8 ✦ MARSH GAS ✦ METHYL HYDRIDE

METHIONINE

Products and Uses: An amino acid essential in human nutrition. A dietary supplement and nutrient. An additive in a cosmetic cream and lotion texturizer. Used as a vegetable oil enricher and toiletry conditioner.

Precautions: Mildly toxic by swallowing. A possible human mutagen (changes inherited characteristics).

Synonyms: CAS: 63-68-3 ✦ CYMETHION ✦ LIQUIMETH

METHOXSALEN

Products and Uses: Used in skin products for suntan lotions, gels, mousses, creams, and ointments. It is a suntan accelerator and sunburn protector.

Precautions: A confirmed carcinogen (causes cancer). Moderately toxic by swallowing. A human mutagen (changes inherited characteristics).

Synonyms: CAS: 298-81-7 ✦ XANTHOTOXIN ✦ AMMOIDIN ✦ MELADININ ✦ MELOXINE ✦ METHOXA-DOME ✦ 8-METHOXYPSORALEN ✦ OXSORALEN ✦ OXYPSORALEN ✦ PRORALONE-MOP

METHYL ABIETATE

Products and Uses: A masticatory substance in chewing gum base, in lacquers, varnishes, adhesives, and coatings as a solvent, softener, and plasticizer.

Precautions: A skin irritant. Probably slightly toxic.

Synonyms: CAS: 127-25-3 ✦ ABIETIC ACID, METHYL ESTER ✦ METHYL ESTER OF WOOD ROSIN ✦ METHYL ESTER OF WOOD ROSIN, PARTIALLY HYDROGENATED (FCC)

METHYL ALCOHOL

Products and Uses: In antifreeze, solvents, shellacs, dyes, utility plant fuel, and home heating oil as deicer, solvent, and oil extender.

Precautions: A human poison by swallowing. Mildly toxic by breathing. Effects on the human body are changes in circulation, cough, dyspnea (shortness of breath), headache, eyes watering, nausea, vomiting, blindness, and respiratory effects. A narcotic. A human mutagen (changes inherited characteristics). Flammable, dangerous fire risk. A cumulative poison.

Synonyms: CAS: 67-56-1 ✦ METHANOL ✦ WOOD ALCOHOL ✦ CARBINOL ✦ METHYL HYDROXIDE ✦ PYROXYLIC SPIRIT ✦ WOOD NAPHTHA ✦ METHYLOL ✦ WOOD SPIRIT

METHYLAMINE

Products and Uses: Found in dyes, pharmaceuticals, fuel additives, insecticides, and fungicides.

Precautions: Moderately toxic by breathing. A severe skin irritant. A human mutagen (changes inherited characteristics).

Synonyms: CAS: 74-89-5 ✦ MONOMETHYLAMINE ✦ AMINOMETHANE ✦ CARBINAMINE ✦ MERCURIALIN

METHYL AMYL KETONE

Products and Uses: In perfumery it produces a peppery-fruit odor. Also used in lacquers, solvents, and as a flavoring agent.

Precautions: Moderately toxic by swallowing. Mildly toxic by breathing and skin contact. A skin irritant.

Synonyms: CAS: 110-43-0 ✦ n-AMYL METHYL KETONE ✦ AMYL METHYL KETONE ✦ 2-HEPTANONE ✦ METHYL-AMYL-CETONE ✦ METHYL AMYL KETONE ✦ METHYL PENTYL KETONE

METHYL ANTHRANILATE

Products and Uses: It is a grapelike, fruity flavoring, an odorant in skin cosmetics and hair pomades. Strangely, it is a bird deterrent and used near ponds and lakes where excessive numbers of Canadian geese are a problem.

Precautions: Moderately toxic by swallowing. A skin irritant. Could cause allergic reactions in susceptible individuals.

Synonyms: CAS: 134-20-3 ✦ METHYL-o-AMINOBENZOATE ✦ NEROLI OIL, ARTIFICIAL

METHYLBENZETHONIUM CHLORIDE

Products and Uses: Useful as a bactericide for external skin lesions and abrasions. It is a germ-killing medication.

Precautions: Mildly toxic by swallowing.

Synonyms: CAS: 25155-18-4 ✦ BACTINE

METHYL BROMIDE

Products and Uses: A powerful fumigant gas used to fumigate homes and other buildings under tenting. Also used to destroy infestations of insects in fruits and vegetables.

Precautions: A suspected carcinogen (may cause cancer). Extremely poisonous by breathing. Breathing also causes anorexia, nausea, and vomiting. Corrosive to the skin; can produce severe burns in liquid form. A human mutagen (changes inherited characteristics).The effects are cumulative and damaging to the nervous system, kidneys, and lungs. Central nervous system effects include blurred vision, mental confusion, numbness, tremors, and speech defects.

Synonyms: CAS: 74-83-9 ✦ BROMOMETHANE ✦ DOWFUME ✦ HALON 1001 ✦ ISCOBROME ✦ METAFUME ✦ METAGAS ✦ METHOGAS ✦ PROFUME

METHYL CAPROATE

Products and Uses: Frequently found in skin creams, detergents, and lubricants. Used as a stabilizer and emulsifier (maintains mixes to aid in suspension of oily liquids).

Precautions: Harmless when used for intended purposes.

Synonyms: CAS: 106-70-7 ✦ METHYL HEXANOATE ✦ METHYL ESTER CAPROIC ACID

METHYL CELLULOSE

Products and Uses: Common ingredient in adhesives and paint pigments. Also used in baked goods, fruit pie fillings, meat patties, vegetable patties, and diet foods as a binder, bodying agent, bulking agent, emulsifier (aids in suspension of oily liquids), film former, stabilizer, thickening, and sizing agent.

Precautions: GRAS (generally recognized as safe) when used within FDA limitations.

Synonyms: CAS: 9004-67-5 ✦ CELLULOSE METHYL ETHER ✦ METHOCEL

METHYLENE CHLORIDE

Products and Uses: Used in coffee (decaffeinated), fruits, hops extract, spice oleoresins, vegetables, adhesives, glues, cleaners, waxes, oven cleaners, paint strippers, paint removers, shoe polish, varnishes, stains, and sealants. It is also a degreasing and cleaning fluid used as a solvent for food processing. Used for color dye or fixative and as an extraction solvent. Not used much as a propellant anymore.

Precautions: A confirmed carcinogen (causes cancer). Moderately toxic by swallowing. Mildly toxic by breathing. Effects on the human body by swallowing and breathing are paresthesia (abnormal sensation of burning or tingling), somnolence (sleepiness), altered sleep time, convulsions, euphoria, and change in cardiac rate. An eye and severe skin irritant. A human mutagen (changes inherited characteristics). FDA approves use within limitations.

Synonyms: CAS: 75-09-2 ✦ AEROTHENE MM ✦ DCM ✦ DICHLOROMETHANE ✦ FREON 30 ✦ METHANE DICHLORIDE ✦ METHYLENE BICHLORIDE ✦ METHYLENE DICHLORIDE ✦ SOLMETHINE

METHYL ETHYL CELLULOSE

Products and Uses: An ingredient in meringues and whipped toppings. Pharmaceutically used in laxative products. It is an emulsifier, foaming agent, stabilizer, aerator, and bowel-bulking agent.

Precautions: FDA approves use at moderate levels to accomplish the intended effect.

Synonyms: CELLULOSE, MODIFIED

METHYL ETHYL KETONE

Products and Uses: Common ingredient in paint removers, cements, and cleaning fluids. Used as a solvent, adhesive, surface coating, and stain remover.

Precautions: Moderately toxic by swallowing and skin contact. Effects on the human body by breathing: nose, throat, conjunctiva (around eye) irritation, and unspecified effects on the respiratory system. A strong irritant. Affects peripheral nervous system and central nervous system.

Synonyms: CAS: 78-93-3 ✦ 2-BUTANONE ✦ ETHYL METHYL KETONE ✦ MEK ✦ METHYL ACETONE

METHYL GLYCOL

Products and Uses: Very versatile chemical used in perfumes, colors, soft drink syrups, flavoring extracts, cleansing creams, fabric softeners, suntan lotions, brake fluids, antifreeze, coolants, deicers, and tobacco as a solvent, conditioner, wetting agent, humectant (keeps product from drying out), emulsifier (stabilizes and maintains mixes), anticaking agent, preservative, mold and fungus retarder. Can be used as a spray mist to disinfect the air.

Precautions: Slightly toxic by swallowing and skin contact. Effects on the human body by swallowing are convulsions and general anesthesia. An eye and skin irritant.

Synonyms: CAS: 57-55-6 ✦ 1,2-PROPYLENE GLYCOL ✦ 1,2-DIHYDROXYPROPANE ✦ 1,2-PROPANEDIOL ✦ METHYLENE GLYCOL ✦ POLYPROPYLENE GLYCOL

METHYLHEPTENONE

Products and Uses: Found in inexpensive perfumes and odorants for citrus-lemon fragrance and taste.

Precautions: Moderately toxic by swallowing. A skin irritant.

Synonyms: CAS: 409-02-9 ✦ 6-METHYL-5-HEPTENE-2-ONE

METHYLOACRYLAMIDE

Products and Uses: For coating, cementing, and textile treatment in varnishes, adhesives, crease-proof fabrics, wrinkle-resistant fabrics, and permanent press textiles.

Precautions: Harmless when used for intended purposes. Some individuals experience allergic skin reactions when they come in contact with these materials.

Synonyms: NONE FOUND.

METHYLPARABEN

Products and Uses: A common preservative in cosmetics, facial moisturizers, lipstick, eye shadow, mascara, and others. It is an antimicrobial (germ killing) agent to prevent bacterial growth in makeup products.

Precautions: Could cause allergic reaction. Moderately toxic by swallowing. A mutagen (changes inherited characteristics).

Synonyms: CAS: 99-76-3 ✦ PARABENS ✦ METHYL-p-HYDROXYBENZOATE ✦ ABIOL ✦ ASEPTOFORM ✦ METHYLBEN ✦ METHYL CHEMOSEPT ✦ METHYL PARASEPT ✦ METOXYDE ✦ MOLDEX ✦ PARASEPT ✦ PARIDOL ✦ SEPTOS ✦ PRESERVAL M ✦ TEGOSEPT M

METHYL PHENYLACETATE

Products and Uses: Honeylike odor for perfumery, flavors for tobacco, and flavoring. Ingredient to affect taste and/or smell of products.

Precautions: Could cause allergic reaction in susceptible individuals.

Synonyms: NONE FOUND.

METHYLPHOSPHORIC ACID

Products and Uses: Ingredient in rust remover. Used for steel finishes on vehicles, farm equipment, bicycles, toys, and so on.

Precautions: Directions must be closely followed on products that contain this chemical.

Synonyms: METHYL ORTHOPHOSPHORIC ACID ✦ METHYL ACID PHOSPHATE

METHYL SALICYLATE

Products and Uses: Used as a UV (ultaviolet) absorber in sunburn lotions. It is a flavoring in foods, baked goods, beverages, chewing gum, candy, odorant in perfumery, and a topical analgesic (pain killer).

Precautions: Human poison by swallowing. Effects in the human body by swallowing are dyspnea (shortness of breath), nausea, vomiting, and respiratory stimulation. A severe skin and eye irritant. Swallowing of relatively small amounts has caused severe poisoning and death.

Synonyms: CAS: 119-36-8 ✦ o-ANISIC ACID ✦ BETULA OIL ✦ GAULTHERIA OIL, ARTIFICIAL ✦ o-HYDROXYBENZOIC ACID, METHYL ESTER ✦ 2-HYDROXYBENZOIC ACID METHYL ESTER ✦ o-METHOXYBENZOIC ACID ✦ 2-METHOXYBENZOIC ACID ✦ METHYL-o-HYDROXYBENZOATE ✦ NATURAL WINTERGREEN OIL ✦ OIL of WINTERGREEN ✦ SALICYLIC ACID, METHYL ESTER ✦ SWEET BIRCH OIL ✦ SYNTHETIC WINTERGREEN OIL ✦ TEABERRY OIL ✦ WINTERGREEN OIL ✦ WINTERGREEN OIL, SYNTHETIC

β-METHYLUMBELLIFERONE

Products and Uses: An ingredient in soaps, starches, laundry products, and suntan lotions. It produces a bright, blue-white fluorescence in daylight or UV (ultraviolet) light. Causes white fabrics to appear brighter and cleaner.

Precautions: Could cause allergic reaction on skin contact with treated materials.

Synonyms: BMU ✦ 7-HYDROXY-4-METHYL COUMARIN

MILK OF MAGNESIA

Products and Uses: A common laxative medication.

Precautions: Harmless when used for intended purposes and package directions are followed.

Synonyms: MAGNESIA MAGMA ✦ MAGNESIUM HYDROXIDE

MINERAL OIL

Products and Uses: Used in bakery products, beet sugar, confectionery, egg white solids, fruit (raw), fruits (dehydrated), grain, meat (frozen), pickles, potatoes (sliced), sorbic acid, starch (molding), vegetables (dehydrated), vegetables (raw), vinegar, wine, and yeast. It is also a laxative. Used in cosmetic creams, baby lotions, hair conditioners, lipstick, mascara, blush, shaving creams, makeup foundations, cleansing creams, and suntan lotions. It is a binder, defoaming agent, fermentation aid, lubricant, coating (protective), and release agent.

Precautions: A petroleum derivative. A human teratogen (abnormal fetal development) by breathing which causes testicular tumors in the fetus. Breathing of vapor or particulates can cause aspiration pneumonia. A skin and eye irritant. A possible carcinogen (causes cancer) producing gastrointestinal (stomach and intestines) tumors. Combustible liquid when exposed to heat or flame. There is also a purified food grade which is approved for use within limitations by the FDA.

Synonyms: CAS: 8012-95-1 ✦ ADEPSINE OIL ✦ ALBOLINE ✦ BAYOL F ✦ BLANDLUBE ✦ CRYSTOSOL ✦ DRAKEOL ✦ FONOLINE ✦ GLYMOL ✦ KAYDOL ✦ KONDREMUL ✦ MINERAL OIL, WHITE ✦ MOLOL ✦ NEO-CULTOL ✦ NUJOL ✦ OIL MIST, MINERAL ✦ PAROL ✦ PAROLEINE ✦ PARRAFIN OIL ✦ PENETECK ✦ PENRECO ✦ PERFECTA ✦ PETROGALAR ✦ PETROLATUM, LIQUID ✦ PRIMOL 335 ✦ PROTOPET ✦ SAXOL ✦ TECH PET F ✦ WHITE MINERAL OIL

MONAMINES

Products and Uses: In detergents, detergent additives, soaps, shampoos, cleansers, and cleaning agents. Used as foam boosters, wetters, emulsifiers (breaks down soil and grease), dispersing agents (mixes and spreads ingredients), thickeners, and conditioners.

Precautions: Could cause allergic reaction in susceptible individuals.

Synonyms: DIALKYLOLAMIDES

MONOGLYCERIDE

Products and Uses: Used in cosmetics, creams, lotions, facial rouges, mascara, and eye shadow as an emulsifier (stabilizes and maintains mixes to aid in suspension of oily liquids), and lubricant.

Precautions: Could cause allergic reaction in susceptible individuals.

Synonyms: GLYCEROL MONOSTEARATE   ✦   GLYCEROL MONOLAURATE

MONOSODIUM GLUTAMATE

Products and Uses: Used in meat, poultry, sauces, soups, pickles, condiments, bakery products, and candies to enhance the taste of foods. Occurs naturally in seaweed.

Precautions: Mildly toxic by swallowing. Effects on the body by ingesting are somnolence (sleepiness), hallucinations, distorted perceptions, headache, dyspnea (shortness of breath), nausea, vomiting, and dermatitis. The cause of "Chinese restaurant syndrome." Some people experience mood changes, irritability, numbness, and depression. It can trigger IBS (irritable bowel syndrome). Animal studies resulted in reproductive effects (infertility, or sterility, or birth defects) and teratogenic (abnormal fetal development) effects. FDA permits use within limitations. It has a pharmacologic effect. This means that given enough MSG anyone would develop symptoms. This is different than an allergic effect. It could be compared to the effect of alcohol on individuals. People have different tolerances to it.

Synonyms: CAS: 142-47-2   ✦   ACCENT   ✦   AJINOMOTO   ✦   CHINESE SEASONING   ✦   GLUTACYL   ✦   GLUTAMIC ACID, SODIUM SALT   ✦   GLUTAVENE   ✦   MONOSODIUM-*l*-GLUTAMATE   ✦ α-MONOSODIUM GLUTAMATE   ✦   MSG   ✦   RL-50   ✦   SODIUM GLUTAMATE   ✦   SODIUM *l*-GLUTAMATE   ✦   *l*-(+) SODIUM GLUTAMATE   ✦   VETSIN   ✦   ZEST

MONTAN WAX

Products and Uses: Used in shoe polish, furniture polish, roof paints, waterproof paints, adhesives, pastes, carbon paper, and substitute for carnauba and beeswax as a protective ingredient and decorative coating. It has been reported that sales are down on wax because of low carbon paper sales due to the high use of copying machines.

Precautions: Harmless when used for intended purposes.

Synonyms: LIGNITE WAX

MORPHINE

Products and Uses: Analgesic used to reduce or prevent mortal pain. It is a sedative that reduces anxiety.

Precautions: Narcotic, a habit-forming drug. Sale is restricted by law in the U.S. Poison by swallowing, by subcutaneous (injected under the skin), and intravenous (injected in the vein) use. Causes reproductive (infertility, sterility, or birth defects) effects.

Synonyms: CAS: 57-27-2 ✦ MORPHIA ✦ MORPHINA ✦ MORPHINIUM ✦ MORPHIUM

MORPHOLINE

Products and Uses: An ingredient in waxes, polishes, detergent brightener, pharmaceuticals, bactericides, local anesthetics, and antiseptics.

Precautions: Moderately toxic by swallowing, breathing, and skin contact. A human mutagen (changes inherited characteristics). A corrosive irritant to skin, eyes, nose, and throat. Can cause kidney damage. A possible carcinogen (causes cancer).

Synonyms: CAS: 110-91-8 ✦ TETRAHYDRO-1,4-OXAZINE ✦ DIETHYLENEIMIDE OXIDE ✦ DIETHYLENE IMIDOXIDE ✦ DIETHYLENE OXIMIDE

MUCILAGE

Products and Uses: Glues, adhesives, and gum used for adhesion. They are derived from seeds, roots, plants, algae, or seaweed.

Precautions: Harmless when used for intended purposes.

Synonyms: ALGAL POLYSACCHARIDES ✦ AGAR MUCILAGE ✦ ALGIN MUCILAGE ✦ CARRAGEENIN MUCILAGE

MUSK

Products and Uses: A popular additive in cosmetics, shaving lotions, shaving creams, colognes, toilet waters, air fresheners, and mothproofers. Useful as a fragrance or odorant.

Precautions: Can cause allergic reaction in susceptible individuals. Not intended for consumption.

Synonyms: CAS: 300-54-9 ✦ MUSCARINE ✦ MUSCARIN ✦ MUSKARIN ✦ *dl*-MUSCARINE

MYRISTIC ACID

Products and Uses: An additive in cosmetic, soaps, and shaving creams as a defoaming agent and lubricant.

Precautions: A human mutagen (changes inherited characteristics). A human skin irritant. Could cause allergic reactions.

Synonyms: CAS: 544-63-8 ✦ CRODACID ✦ EMERY 655 ✦ HYDROFOL ACID 1495 ✦ HYSTRENE 9014 ✦ TETRADECANOIC ACID ✦ n-TETRADECOIC ACID ✦ 1-TRIDECANECARBOXYLIC ACID ✦ UNIVOL U 316S

MYRISTYL ALCOHOL

Products and Uses: A component in soaps, cosmetics, detergents, ointments, suppositories, shampoo, toothpaste, cold cream, and cleaning preparations as a fixative, wetting agent, and antifoam agent.

Precautions: A human skin irritant.

Synonyms: CAS: 112-72-1 ✦ 1-TETRADECANOL ✦ TETRADECYL ALCOHOL

MYRRH OIL

Products and Uses: Found in incense, perfumes, dentifrices, foods, and beverages as an ingredient to affect the taste or smell of products.

Precautions: Harmless when used for appropriate purposes. FDA approves use at moderate levels to accomplish the intended effect.

Synonyms: CAS: 9000-45-7

MYXIN

Products and Uses: A germicide or fungicide used as an antibiotic, bacteriostat, and antifungal.

Precautions: Harmless when used for intended effect.

Synonyms: CAS: 13925-12-7 ✦ 6-METHOXY-1-PHENAZINOL-5,10-DIOXID

NAPHTHA

Products and Uses: Found in paints, stains, finishes, varnish thinner, dry-cleaning fluid, asphalt solvent, and naphtha soaps. Used for eggs (shell), fruits (fresh), and vegetables (fresh) as a coloring agent, coating (protective), solvent, and various other uses.

Precautions: Mildly toxic by breathing. Can cause unconsciousness which may go into coma, labored breathing, and bluish tint to the skin. Recovery follows removal from exposure. In mild form, intoxication resembles drunkenness. On a chronic basis, no true poisoning; sometimes headache, lack of appetite, dizziness, sleeplessness, indigestion, and nausea. Flammable when exposed to heat or flame.

Synonyms: CAS: 8030-30-6 ✦ AROMATIC SOLVENT ✦ BENZIN ✦ COAL TAR NAPHTHA ✦ HI-FLASH NAPHTHAETHYLEN ✦ NAPHTA (DOT) ✦ NAPHTHA DISTILLATE ✦ NAPHTHA PETROLEUM ✦ NAPHTHA, SOLVENT ✦ PETROLEUM BENZIN ✦ PETROLEUM DISTILLATES ✦ PETROLEUM ETHER ✦ PETROLEUM NAPHTHA ✦ PETROLEUM SPIRIT ✦ SKELLY-SOLVE-F ✦ VM&P NAPHTHA

NAPHTHALENE

Products and Uses: Commonly used for moth balls, rug cleaner, upholstery cleaner, preservative, and antiseptic; a squirrel and small animal repellent.

Precautions: Human poison by swallowing. An eye and skin irritant. Can cause nausea, headache, diaphoresis (perspiration), hematuria (blood in urine), fever, anemia, liver damage, vomiting, convulsions, and coma. Poison by swallowing, breathing, or skin absorption.

Synonyms: CAS: 91-20-3 ✦ TAR CAMPHOR ✦ MOTH BALLS ✦ MOTH FLAKES ✦ NAPHTHALIN ✦ NAPHTHENE ✦ WHITE TAR ✦ CAMPHOR TAR

β-NAPHTHOL

Products and Uses: An additive in pigments, dyes, insecticides, pharmaceuticals, perfumes, and antiseptics. Used as a coloring agent, preservative, pesticide, and germ killer.

Precautions: Poison by swallowing. A skin and eye irritant.

Synonyms: CAS: 135-19-3 ✦ 2-NAPHTHOL ✦ 2-HYDROXYNAPHTHALENE ✦ ISONAPHTHOL ✦ 2-NAPHTHALENOL ✦ DEVELOPER SODIUM ✦ DEVELOPER AMS ✦ β-MONOOXYNAPHTHALENE ✦ β-NAPHTHYL ALCOHOL ✦ β-NAPHTHYL HYDROXIDE

β-NAPHTHYL ETHYL ETHER

Products and Uses: A component in perfumes, soaps, toiletries, and cosmetics. An odorant or fragrance with orange blossom aroma.

Precautions: Could cause allergic reaction in susceptible individuals.

Synonyms: NEROLIN

NARCOTIC

Products and Uses: Any drug capable of producing anesthesia and analgesia. An addictive drug. There are natural (morphine, codeine) and synthetic (meperidine, ethadone, and phenazocine) forms.

Precautions: The sale of narcotics is strictly controlled by law in the U.S. These substances aid sleep and pain relief, but may result in addiction (a situation where the body cannot function normally).

Synonyms: OPIUM ✦ HEROIN ✦ DIONIN

NARINGIN

Products and Uses: Used as a sweetener in beverages.

Precautions: Harmless when used for intended purposes.

Synonyms: CAS: 10236-47-2 ✦ NARINGENIN-7-RHAMNOGLUCOSIDE ✦ NARINGENIN-7-RUTINOSIDE ✦ AURANTIIN

NEATSFOOT OIL

Products and Uses: A polishing, lubricating, preserving, and finishing agent for leather and wool items.

Precautions: A combustible. Label instructions must be followed.

Synonyms: NONE KNOWN.

NEOPRENE

Products and Uses: Used for coating, sealing, waterproofing, adhering materials, in cements, carpet backings, sealants, gaskets, and adhesive tapes. Produced in both liquid and foam.

Precautions: Poison by swallowing. Moderately toxic by breathing. Can cause dermatitis, eye inflammation, corneal necrosis, anemia, hair loss, nervousness, and irritability,

Synonyms: CAS: 126-99-8 ✦ POLYCHLOROPRENE ✦ CHLOROBUTADIENE ✦ CHLOROPRENE

NICKEL

Products and Uses: Used for coins, batteries, electrical components, alloys, protective coatings, and also the hydrogenation of vegetable oils.

Precautions: A confirmed carcinogen. Poison by swallowing. Swallowing of soluble salts causes nausea, vomiting, and diarrhea. A mutagen (changes inherited characteristics). Hypersensitivity to nickel is common and can cause allergic contact dermatitis, asthma, conjunctivitis (inflammation of tissue surrounding eye), inflammatory reactions around nickel-containing medical implants and prostheses.

Synonyms: CAS: 7440-02-0 ✦ Ni 270 ✦ NICKEL 270 ✦ NICKEL (DUST) ✦ NICKEL SPONGE ✦ Ni 0901-S ✦ Ni 4303T ✦ RANEY ALLOY ✦ RANEY NICKEL

NICOTINE

Products and Uses: Found in tobacco cigarettes, cigars, pipe tobacco, tobacco products, and smokeless tobacco. Also used in pesticides and in veterinary medicine as an external parasiticide (kills parasites), fumigant, and stimulant.

Precautions: An alkoloid (group of nitrogenous organic compounds, mostly used as pain relievers such as cocaine, quinine, caffeine) from tobacco. A deadly human poison. A human teratogen (abnormal fetal development) by swallowing, causes developmental abnormalities of the cardiovascular system. Causes blood pressure effects. Can be absorbed by intact skin.

Synonyms: CAS: 54-11-5 ✦ BLACK LEAF ✦ NICOCIDE ✦ NICOTINE ALKALOID ✦ ORTH N-4 DUST ✦ XL ALL INSECTICIDE

NIGROSINE

Products and Uses: An additive in ink, shoe polish, leather dyes, wood stains, and textile dyes; also useful as a shark repellent.

Precautions: Harmless when used for intended purposes.

Synonyms: NONE KNOWN.

NISIN PREPARATION

Products and Uses: A component included as a preservative and antimicrobial agent in cheese spreads (pasteurized), canned fruits, and vegetables.

Precautions: Harmless when used for intended purposes within FDA limitations.

Synonyms: CAS: 1414-45-5 ✦ STREPTOCOCCUS LACTUS

NITRATES

Products and Uses: Used in matches, cigars, cigarettes, and tobacco products to keep the tobacco burning evenly. Banned as a food preservative.

Precautions: When combined with saliva and secondary amines (food substances), nitrosamines (cancer-causing agents) are formed.

Synonyms: CAS: 7757-79-1 ✦ POTASSIUM NITRATE ✦ SODIUM NITRATE

NITRITES

Products and Uses: A curing agent, color fixative, and flavor preserver in bacon, meat (cured), meat products, smoked fish, frankfurters, bologna, and poultry products. Prevents growth of botulism spores.

Precautions: Large amounts taken by mouth may produce nausea, vomiting, cyanosis (bluish skin color), collapse, and coma. Repeated small doses cause a fall in blood pressure, rapid pulse, headache, and visual disturbances. They have been implicated in an increased incidence of cancer. FDA approves use to accomplish the effect when used within stated limits. Nitrosamines (cancer-causing agents) are formed when nitrites are combined with stomach and food chemicals. FDA states that the addition of Vitamin C reduces the formation of nitrosamines.

Synonyms: POTASSIUM NITRITE ✦ SODIUM NITRITE

NITROBENZENE

Products and Uses: An additive in soaps, cleaning products, metal polish, shoe polish, floor and furniture polishes.

Precautions: Moderately toxic by swallowing and skin contact. Effects on the human body by swallowing include anesthesia, respiratory (breathing) stimulation and vascular (blood vessel) changes. An eye and skin irritant. Absorbed rapidly through the skin. The vapors are hazardous.

Synonyms: CAS: 98-95-3 ✦ OIL OF MIRBANE ✦ ESSENCE OF MIRBANE ✦ OIL OF MYRBANE ✦ MIRBANE OIL

NITROCELLULOSE

Products and Uses: Various applications include as a propellant in small arms ammunition, explosives, vehicle paint lacquers, printing ink, and celluloid.

Precautions: Dangerous fire and explosion risk.

Synonyms: CAS: 9004-70-0 ✦ CELLULOSE NITRATE ✦ NITROCOTTON ✦ GUNCOTTON ✦ PYROXYLIN

NITROGEN

Products and Uses: Used as a food antioxidant (slows down spoiling due to oxygen) for fruit, poultry, wine, and various food in sealed containers, also in-transit food refrigeration and freeze drying. Useful as an aerating agent, a gas, modified atmospheres for insect control (truck or container is filled with nitrogen to kill insects or rodents), oxygen exclusion, and propellant.

Precautions: Low toxicity. In high concentrations it is a simple asphyxiant. The release of nitrogen from solution in the blood, with formation of small bubbles, is the cause of most of the symptoms and changes found in compressed air illness (caisson disease, decompression disease, and the bends). It is a narcotic at high concentration and high pressure. Both the narcotic effects and the bends are hazards of compressed air atmospheres such as found in underwater diving. Nonflammable gas.

Synonyms: CAS: 7727-37-9 ✦ NITROGEN, COMPRESSED ✦ NITROGEN, REFRIGERATED LIQUID ✦ NITROGEN GAS

NITROGEN OXIDE

Products and Uses: Utilized as an aerating agent, gas, anesthetic in dentistry and surgery; propellant for food and cosmetic products. Also used in simulated dairy products (whipped creams) and in wine.

Precautions: Moderately toxic by breathing. Effects on the human body by breathing are general anesthetic, decreased pulse rate without blood pressure fall, and body temperature decrease. An asphyxiant. Moderate explosion hazard.

Synonyms: CAS: 10024-97-2 ✦ DINITROGEN MONOXIDE ✦ FACTITIOUS AIR ✦ HYPONITROUS ACID ANHYDRIDE ✦ LAUGHING GAS ✦ NITROUS OXIDE ✦ NITROUS OXIDE, COMPRESSED ✦ NITROUS OXIDE, REFRIGERATED LIQUID

NONYL PHENOL

Products and Uses: Various uses include solvent-type products in detergents, emulsifiers, stain removers, oil additives, fungicides, and preservatives for plastics and rubbers.

Precautions: Moderately toxic by swallowing and skin contact.

Synonyms: CAS: 25154-52-3 ✦ ISOMERIC MONOALKYL PHENOL MIXTURE

NORMAL PARAFFIN SOLVENTS

Products and Uses: Active component in waterless hand cleaner or solvent formulation.

Precautions: Harmless when used for intended purposes.

Synonyms: NORPAR

OCHER

Products and Uses: Various colored earthy powders in paint pigments, stains and inks, wallpaper pigment, artists' colors, cosmetics, and theatrical makeup.

Precautions: Harmless when used for intended purposes.

Synonyms: UMBER ✦ SIENNA ✦ HYDRATED FERRIC OXIDE MIXTURES ✦ CALCINED (BURNT OCHER)

γ-OCTALACTONE

Products and Uses: A peach-flavored additive in baked goods, candy, or ice cream.

Precautions: Mildly toxic by swallowing. A skin irritant. FDA approves use at a moderate level to accomplish the intended effect.

Synonyms: CAS: 104-50-7 ✦ γ-n-BUTYL-γ-BUTYROLACTONE ✦ 5-HYDROXYOCTANOIC ACID LACTONE ✦ OCTANOLIDE-1,4 ✦ TETRAHYDRO-6-PROPYL-2H-PYRAN-2-ONE

1-OCTANAL

Products and Uses: An odorant in soaps, cosmetics, perfumes, and toiletry products.

Precautions: Mildly toxic by swallowing and skin contact. A skin and eye irritant. FDA approves use at moderate levels to accomplish the intended effect.

Synonyms: CAS: 124-13-0 ✦ ALDEHYDE C-8 ✦ C-8 ALDEHYDE ✦ FEMA NO. 2797 ✦ OCTANALDEHYDE ✦ n-OCTYL ALDEHYDE

OCTANE

Products and Uses: An antiknocking agent in internal combustion engines.

Precautions: May act as a simple asphyxiant. A narcotic in high concentrations. Extended skin contact can cause blisters. Brief skin contact causes burning sensation. A dangerous fire and explosion hazard.

Synonyms: CAS: 111-65-9 ✦ OKTAN ✦ OKTANEN ✦ OTTANI

OCTHILINONE

Products and Uses: A mildewcide (destroys mold and mildew), fungicide, and biocide (kills bacteria) in paints, cosmetics, and shampoos.

Precautions: Moderately toxic by swallowing and skin contact. A skin and severe eye contact.

Synonyms: CAS: 26530-20-1 ✦ KATHON LP PRESERVATIVE ✦ MICROCHEK ✦ PANCIL ✦ SKANE

OCTYL ACETATE

Products and Uses: An odorant in perfumed cosmetics, hair products, skin lotions, and fragrances.

Precautions: Moderately toxic by swallowing. A skin and eye irritant.

Synonyms: CAS: 103-09-3 ✦ ACETATE C-8 ✦ CAPRYLYL ACETATE ✦ 2 ETHYLHEXANYL ACETATE ✦ 2 ETHYLHEXYL ACETATE

OCTYL ALCOHOL

Products and Uses: An additive in colognes and cosmetics; used as an antifoaming, scenting, and flavoring agent. Used in beverages, candy, gelatin desserts, ice cream, and pudding mixes.

Precautions: Moderately toxic by swallowing. A mutagen (changes inherited characteristics). A skin irritant. FDA approves use only for encapsulating lemon, lime, orange, peppermint, and spearmint oil in limited levels.

Synonyms: CAS: 111-87-5 ✦ ALCOHOL C-8 ✦ ALFOL 8 ✦ CAPRYL ALCOHOL ✦ CAPRYLIC ALCOHOL ✦ DYTOL M-83 ✦ EPAL 8 ✦ HEPTYL CARBINOL ✦ 1-HYDROXYOCTANE ✦ LOROL 20 ✦ OCTANOL ✦ n-OCTANOL ✦ 1-OCTANOL ✦ OCTILIN ✦ OCTYL ALCOHOL, NORMAL-PRIMARY ✦ PRIMARY OCTYL ALCOHOL ✦ SIPOL L8

OIL OF CARDAMOM

Products and Uses: A seasoning agent in liqueurs, cold medications, cough syrups, sauces, candies, and bakery products.

Precautions: Could cause allergic reaction in susceptible individuals.

Synonyms: *ELETTARIA CARDAMOMUM*

OIL OF LIME, distilled

Products and Uses: An ingredient used to affect the taste or smell of bakery products, beverages (nonalcoholic), chewing gum, condiments, confections, gelatin desserts, ice cream products, and puddings.

Precautions: A skin irritant. A possible carcinogen which caused tumors in laboratory animals. A mutagen (changes inherited characteristics). FDA states GRAS (generally recognized as safe) when used within reasonable limits.

Synonyms: CAS: 8008-26-2 ✦ DISTILLED LIME OIL ✦ LIME OIL ✦ LIME OIL, DISTILLED ✦ OILS, LIME

OIL OF MACE

Products and Uses: Utilized as a flavoring agent in bread, cakes, chocolate pudding, and fruit salad.

Precautions: Moderately toxic by swallowing. A skin irritant. FDA states GRAS (generally recognized as safe) when used at moderate levels to accomplish the intended effects.

Synonyms: CAS: 8007-12-3 ✦ NCI-C56484 ✦ MACE OIL ✦ OIL OF NUTMEG, EXPRESSED

OIL OF ORANGE

Products and Uses: Applications include seasoning in bakery products, beverages (nonalcoholic), chewing gum, condiments, ice cream products, cough syrup, and cold medications

Precautions: A skin irritant. FDA states GRAS (generally recognized as safe) when used within reasonable limits.

Synonyms: CAS: 8008-57-9 ✦ NEAT OIL OF SWEET ORANGE ✦ OIL OF SWEET ORANGE ✦ ORANGE OIL ✦ ORANGE OIL, COLDPRESSED ✦ SWEET ORANGE OIL

OLEIC ACID

Products and Uses: A chemical found in beet sugar, citrus fruit (fresh), sugar beets, yeast, soaps, ointments, cosmetics, hair wave products, shaving creams, lipstick, shampoos, liquid makeup, nail polishes, polishes, and waterproofing compounds as a binding, coating, defoaming, and lubricating agent.

Precautions: Mildly toxic by swallowing. A mutagen (changes inherited characteristics). A skin irritant. A possible carcinogen (causes cancer) that caused tumor growth in laboratory animals. FDA approves use within specific limitations.

Synonyms: CAS: 112-80-1 ✦ CENTURY CD FATTY ACID ✦ EMERSOL ✦ EMERSOL 221 LOW TITER WHITE OLEIC ACID ✦ EMERSOL 220 WHITE OLEIC ACID ✦ GLYCON WO ✦ GROCO 2 ✦ HY-PHI 2102 ✦ INDUSTRENE 105 ✦ K 52 ✦ *L'ACIDE OLEIQUE (FRENCH)* ✦ METAUPON ✦ NEO-FAT 90-04 ✦ cis-Δ⁹-OCTADECENOIC ACID ✦ cis-OCTADEC-9-ENOIC ACID ✦ cis-9-OCTADECENOIC ACID ✦ 9,10-OCTADECENOIC ACID ✦ PAMOLYN ✦ RED OIL ✦ TEGO-OLEIC 130 ✦ VOPCOLENE 27 ✦ WECOLINE OO ✦ WOCHEM NO. 320

OLIVE OIL

Products and Uses: Utilized in soaps, textile soaps, cosmetics, skin creams, medical ointments, liniments, emollient, salad dressings, sardine packing, anchovy packing, and laxatives.

Precautions: Harmless when used for intended purposes.

Synonyms: CAS: 8001-25-0 ✦ *OLGA EUROPA* ✦ *OLEACEAE*

ORANGE B

Products and Uses: A coloring ingredient in foods such as frankfurter and sausage casings.

Precautions: FDA approves when used within limitations.

Synonyms: CAS: 15139-76-1 ✦ 1-(4-SULFOPHENYL)-3-ETHYLCARBOXY-4-(4-SULFONAPHTHYLAZO)-5-HYDROXYPYRAZOLE

ORRIS

Products and Uses: A plant root used for fruit and spice flavorings in food and beverages; as an aroma ingredient for toiletries. It is used in dusting powder, tooth powder, dry shampoos, sachets, dry fragrances, and cosmetics. Also added to ice creams, candies, bakery goods, desserts, and toppings.

Precautions: Known to cause allergic reactions in susceptible individuals.

Synonyms: ORRIS ROOT OIL ✦ LOVE ROOT ✦ WHITE FLAG

OXALIC ACID

Products and Uses: Usually added to radiator cleaners, concrete cleaners, rug cleaners, upholstery cleaners, toilet bowl cleaners, ink remover, rust remover, metal polishes, wood cleaners, textile bleaching, and skin bleaching preparations.

Precautions: Toxic by breathing and swallowing. A strong irritant.

Synonyms: CAS: 144-62-7 ✦ ETHANEDIOIC ACID

OXYBENZONE

Products and Uses: A common ingredient in sunscreen lotions, ointments, sprays, oils, liquids, and creams. Used as a UV (ultraviolet) skin protector.

Precautions: Harmless when used for intended purposes. Could cause an allergic reaction.

Synonyms: CAS: 131-57-7 ✦ 4-METHOXY-2-HYDROXYBENZOPHNONE

OXYMETHUREA

Products and Uses: Used for textiles as cotton wrinkleproofing, shrinkproofing, finishing and drying. It is found in wash-and-wear fabrics.

Precautions: Could cause contact allergic dermatitis in susceptible individuals.

Synonyms: METHURAL ✦ N'N'-BIS(HYDROXYMETHYL)UREA

OXYSTEARIN

Products and Uses: A defoaming agent and sequestrant (binds constituents) that affect the final product's flavor or texture. Useful for beet sugar, cooking oil, salad oil, vegetable oils, and yeast. It prevents crystal formation also.

Precautions: FDA approves use when used within limitations

Synonyms: NONE KNOWN.

OZOCERITE

Products and Uses: A general purpose wax, sometimes used in place of carnauba and beeswax. Also used in paints, leather polishes, inks, carbon papers, floor polishes, crayons, waxed paper, cosmetics, lipstick, rouges, ointments, and waxed cloth.

Precautions: Harmless when used for intended purposes.

Synonyms: MINERAL WAX ✦ FOSSIL WAX ✦ OZOKERITE ✦ CERESIN

OZONE

Products and Uses: An antimicrobial agent used for disinfecting, oxidizing, bleaching, deodorizing, and purifying. It is used sometimes for bottled water.

Precautions: A human poison by breathing. Effects on the human body by breathing are visual field changes, lacrimation (watery eyes), headache, decreased pulse rate with fall in blood pressure, dermatitis, cough, dyspnea (shortness of breath), respiratory stimulation, and other pulmonary changes. A human mutagen (changes inherited characteristics). A skin, eye, upper respiratory system, nose, and throat irritant. Can be a safe water disinfectant in low concentration. Note: Depletion of the ozone layer in the stratosphere, which acts as a shield against penetration of UV (ultraviolet) light in the sun's rays, is believed to be caused by light-induced chlorofluorocarbon decomposition resulting from increased use of halocarbon aerosol propellants. Their manufacture and use were prohibited in 1979, except for a few specialized items. Chlorofluorocarbon use is being phased out for all home and automobile air conditioners.

Synonyms: CAS: 10028-15-6 ✦ OZON (POLISH) ✦ TRIATOMIC OXYGEN

PABA

> **Products and Uses:** Added to suntan lotion as an ultraviolet (UV) absorber, sunscreen.
>
> **Precautions:** A possible allergen. Swallowing can cause nausea, vomiting, skin rash, and toxic hepatitis. Moderately toxic.
>
> **Synonyms:** CAS: 150-13-0 ✦ 4-AMINOBENZOIC ACID ✦ ANTI-CHROMOTRICHIA FACTOR ✦ p-AMINOBENZOIC ACID ✦ p-CARBOXYPHENYLAMINE ✦ TRICHOCHROMOGENIC FACTOR ✦ VITAMIN H

PALM OIL

> **Products and Uses:** Frequently found in margarine, shortening, soap, candles, lubricant, cosmetics, ointments, balms, and skin lotions. Used as a coating agent, emulsifying agent (stabilizes and maintains mixes), and texturizing agent (improves textures).
>
> **Precautions:** GRAS (generally recognized as safe) when used for intended purposes.
>
> **Synonyms:** PALM BUTTER ✦ PALM TALLOW

PANTHENOL

> **Products and Uses:** B complex vitamin. In cosmetics, hair shampoos, rinses, emollients (softeners), and dietary supplements. Richest sources are queen bee jelly and liver. A nutrient. Hair and skin conditioning and softening.
>
> **Precautions:** FDA states GRAS (generally recognized as safe) when used for intended purposes.

Synonyms: CAS: 81-13-0 ✦ BEPANTHEN ✦ BEPANTHENE ✦
BEPANTOL ✦ COZYME ✦ DEXPANTHENOL ✦
d-(+)-2,4-DIHYDROXY-N-(3-HYDROXYPROPYL)-3,3-DIMETHYLBUTYRAMIDE ✦
D-P-A INJECTION ✦ ILOPAN ✦ MOTILYN ✦ PANADON ✦
PANTHENOL ✦ PANTOL ✦ PANTOTHENOL ✦ PANTOTHENYL
ALCOHOL ✦ ZENTINIC

PAPAIN

Products and Uses: Derived from papaya. Used for meat, poultry, wine, and medications. Also an additive in tobacco, pharmaceuticals, and cosmetics. Used for chillproofing (prevents protein haze) of beer; an enzyme (digests protein), for meat tenderizing; cereals (preparation of precooked), processing aid, and tissue softening agent. Used as an anthelmintic (deworming agent), and medication for chronic dyspepsia (indigestion).

Precautions: Effects on the human body by swallowing are changes in structure or function of esophagus. An allergen. Can cause allergic reaction. FDA states GRAS (generally recognized as safe) when used within limitations.

Synonyms: CAS: 9001-73-4 ✦ ARBUZ ✦ CAROID ✦ NEMATOLYT ✦
PAPAYOTIN ✦ SUMMETRIN ✦ TROMASIN ✦ VEGETABLE PEPSIN ✦
VELARDON ✦ VERMIZYM

PARADICHLOROBENZENE

Products and Uses: Frequently added to toilet bowl cleaners and moth balls. Useful as a caustic cleanser; an insect repellent.

Precautions: A confirmed carcinogen. Moderately toxic to humans by swallowing. Effects on the body by swallowing include changes to the eyes, lungs, thorax, respiration, and decreased motility or constipation. Can cause liver injury. An eye irritant.

Synonyms: CAS: 106-46-7 ✦ PARACHLOROPHENYL CHLORIDE ✦
DICHLORICIDE ✦ DICHLOROBENZENE-PARA ✦ EVOLA ✦ PARA
CRYSTALS ✦ SANTA-CHLOR

PARAFFIN

Products and Uses: Used in chewing gum, candles, wax paper, adhesive component, crayons, floor polishes, cold cream, eyebrow pencils, lipliner pencils, lipstick, chewing gum, pharmaceutical bases, and wax depilatories. Also for coatings, toiletries, lubricating material, masticatory (chewing) substance, food product sealant (for cheeses, cold cuts), tobacco products packaging, water proofing, floor polishing and glass polishing preparations.

Precautions: A possible carcinogen that caused tumors in laboratory animals. Many paraffin waxes contain carcinogens. FDA approves use within limitations to produce the desired results.

Synonyms: CAS: 8002-74-2 ✦ PARAFFIN WAX ✦ PARAFFIN WAX FUME

PARAFORMALDEHYDE

Products and Uses: Destroys fungus and bacterial growth in maple syrup. Used as insecticide, disinfectant, contraceptive cream, and bactericide.

Precautions: Moderately toxic by swallowing. A severe eye and skin irritant. A mutagen (changes inherited characteristics). FDA approves when used within strict limitations.

Synonyms: CAS: 30525-89-4 ✦ FLO-MOR ✦ FORMAGENE ✦ PARAFORSN ✦ TRIFORMOL ✦ TRIOXYMETHYLENE

PARAQUAT

Products and Uses: Used on animal feed, beef, goat, hops (dried), lamb, mint hay (spent), peanuts, pork, and sunflower seed hulls. Has been sprayed on marijuana crops to prevent use of crop. As defoliant, desiccant (dries up crops), and herbicide.

Precautions: Poison by swallowing. A mutagen (changes inherited characteristics). Causes ulceration of digestive tract, diarrhea, vomiting, renal (kidney) damage, jaundice (yellowing of skin), edema (fluid accumulation), hemorrhage, fibrosis of lung. Death from anoxia (absence of oxygen) may result. FDA approves use within limitations.

Synonyms: CAS: 4685-14-7 ✦ DIMETHYL VIOLOGEN ✦ GRAMOXONE S ✦ METHYL VIOLOGEN (2⁺) ✦ PARAQUAT DICATION

PCB

Products and Uses: Added in electrical transformers, components, and systems. It is used for insulating, heat exchange fluid, and hydraulic fluid.

Precautions: On EPA Extremely Hazardous Substance list. A confirmed carcinogen (causes cancer). Moderately toxic by swallowing. Causes skin irritation. Exposed persons may suffer nausea, vomiting, weight loss, jaundice (yellow coloration of eyes and skin), edema (swelling from fluid retention), and abdominal (stomach) pain. Where liver damage has been severe the patient may suffer coma or death.

Synonyms: CAS: 1336-36-3 ✦ AROCLOR ✦ CHLOPEN ✦ CHLOREXTOL ✦ CHLORINATED BIPHENYLS ✦ CHLORINATED DIPHENYLS ✦ CLOPHEN ✦ PHENOCHLOR ✦ POLYCHLORBIPHENYL ✦ PYRALENE ✦ PYRANOL ✦ SANTOTHERM

PECTIN

Products and Uses: An ingredient in beverages, jams, jellies, and cosmetics. As an emulsifier (stabilizes and maintains mixes), gelling agent, stabilizer (to accomplish uniform consistency), and thickening agent.

Precautions: FDA states GRAS (generally recognized as safe) when used at moderate levels to accomplish the intended effects.

Synonyms: CAS: 9000-69-5

PENTACHLOROPHENOL

Products and Uses: Frequently used on telephone poles, pilings, fences, railroad ties, landscape timbers, playground equipment, outdoor wood products, outdoor furniture, in wood varnish, stain, and sealant. Also added to laundry spray starch.

Precautions: Toxic by swallowing, breathing, and skin absorption.

Synonyms: CAS: 87-86-5 ✦ SODIUM PENTACHLOROPHENATE ✦ CHLOROPHEN ✦ DOWCIDE ✦ FUNGIFEN ✦ PENTACHLOR ✦ PENTASOL ✦ PERMAGARD ✦ SANTOPHEN ✦ THOMPSON'S WOOD FIX

PERCHLOROETHYLENE

Products and Uses: A dry-cleaning solvent, rug and upholstery cleaner, spot remover. Used in cleaners, waxes, metal polish, and as degreasing agent. The cleaning agent in 85% of the 30,000 U.S. dry-cleaning stores. They emit 92,000 tons of perc into the air each year.

Precautions: A probable carcinogen (may cause cancer) on EPA list. Moderately toxic to humans by breathing with the following effects: local anesthetic, conjunctiva (eye) irritation, general anesthesia, hallucinations, distorted perceptions (confusion), coma, and pulmonary (lung) changes. A severe eye and skin irritant. The symptoms of acute intoxication from this material are the result of its effects upon the nervous system. Can cause dermatitis, particularly after repeated or prolonged contact with the skin. People who live near or work at dry cleaners have been exposed to perc levels hundred of times higher than the acceptable guidelines. Newly dry-cleaned clothing should not be stored in children's rooms, as they are more sensitive to toxic substances.

Synonyms: CAS: 127-18-4 ✦ ANKILOSTIN ✦ CARBON BICHLORIDE ✦ CARBON DICHLORIDE ✦ DIDAKENE ✦ DOW-PER ✦ ETHYLENE TETRACHLORIDE ✦ FEDAL-UN ✦ NEMA ✦ PERAWIN ✦ PERCHLOR ✦ PERCHLORETHYLENE ✦ PERCLENE ✦ PERCOSOLVE ✦ PERK ✦ PERKLONE ✦ PERSEC ✦ TETLEN ✦ TETRACAP ✦ TETRACHLOROETHENE ✦ 1,1,2,2-TETRACHLOROETHYLENE ✦ TETRALENO ✦ TETRALEX ✦ TETRAVEC ✦ TETROGUER ✦ TETROPIL

PETROLATUM

Products and Uses: A very common petroleum fraction used in multiple ways including in bakery products, beet sugar, confectionery, egg white solids, fruits (dehydrated), vegetables (dehydrated), and yeast. Also ointment bases, pharmaceuticals, cosmetics, rust preventative, modeling clay, and laxatives. Used as lubricating, polishing agent, coating (protective), release agent, and sealing agent.

Precautions: Harmless when used for intended purposes. FDA approves when used within limitations.

Synonyms: WHITE PETROLATUM ✦ YELLOW PETROLATUM ✦ PARAFFIN JELLY ✦ VASOLIMENT ✦ KREMOLINE ✦ VASOLINE ✦ PURELINE

PHENETHYL ALCOHOL

Products and Uses: Frequently used in synthetic rose oil for soaps and flavors, an antibacterial and preservative.

Precautions: Poison by swallowing. Moderately toxic by skin contact. A skin and eye irritant.

Synonyms: CAS: 60-12-8 ✦ PHENYLETHYL ALCOHOL ✦ 2-PHENYLETHANOL ✦ BENZYL CARBINOL

PHENOL

Products and Uses: Used as a topical anesthetic (skin-numbing agent), also as a disinfectant for toilets, outbuildings, floors, drains, and stables. An active ingredient in germicide, slimicide, and cleansing agents.

Precautions: Toxic by swallowing, breathing, and skin absorption. Strong irritant to tissue.

Synonyms: CAS: 108-95-2 ✦ BENZOPHENOL ✦ CRESOLS ✦ XYLENOLS ✦ RESORCINOL ✦ NAPTHOLS ✦ CARBOLIC ACID ✦ PHENYLIC ACID ✦ BENZOPHENOL ✦ HYDROXYBENZENE

PHENOXYACETIC ACID

Products and Uses: Active ingredient in corn remover plasters, pads, drops, callus removers, and exfoliants.

Precautions: Moderately toxic by swallowing. A mild irritant.

Synonyms: CAS: 122-59-8 ✦ GLYCOLIC ACID PHENOL ETHER ✦ PHENOXYETHANOIC ACID ✦ o-PHENYLGLYCOLIC ACID ✦ PHENYLIUM

PHENYLETHYL ANTHRANILATE

Products and Uses: The grape or orange flavor in beverages, synthetic juices, soft drinks, ice creams, ices, candies, and bakery products.

Precautions: Can cause allergic reaction in susceptible individuals.

Synonyms: CAS: 133-18-6 ✦ 2-PHENYLETHYL ANTHRANILATE

PHENYL SALICYLATE

Products and Uses: An additive in adhesives, waxes, polishes, medications, suntan lotions, and creams. Used as a preservative, external disinfectant, intestinal antiseptic, and UV (ultraviolet) absorber.

Precautions: Toxic by swallowing.

Synonyms: CAS: 118-55-8 ✦ SALOL ✦ 2-HYDROXYBENZOIC ACID PHENYL ESTER

PHLOROGLUCINOL

Products and Uses: A cut flower preservative also used in textile dyes and printing inks.

Precautions: Mildly toxic by swallowing.

Synonyms: CAS: 108-73-6 ✦ DILOSPAN S ✦ PHLOROGLUCIN ✦
1,3,5-BENZENETRIOL ✦ 1,3,5-TRIHYDROXYBENZENE

PHOSPHONE ALKYL AMIDE

Products and Uses: A flame and fire retardant for 100% cotton fabrics, tents, military uniforms, draperies, canvas, denims, and camping equipment.

Precautions: Harmless when used for intended purposes.

Synonyms: PYROVATEX

PHOSPHORIC ACID

Products and Uses: The chemical that gives soft drinks their biting taste. Also used in beverages, cheeses, colas, fats (poultry), lard, margarine, poultry, root beer, detergents, and shortening. It is used in rustproofing products and dental cements. It is an acid, sequestrant (binds constituents that affect the final product's appearance, flavor or texture), increases the preservative effect.

Precautions: Moderately toxic by swallowing and skin contact. A corrosive irritant to eyes, skin, nose, and throat; an irritant by breathing. A strong acid. FDA states GRAS (generally recognized as safe) when used within limitations. Consumes calcium from bone tissue in excessive amounts.

Synonyms: CAS: 7664-38-2 ✦ ACIDE PHOSPHORIQUE (FRENCH) ✦ ACIDO
FOSFORICO (ITALIAN) ✦ FOSFORZUUROPLOSSINGEN (DUTCH) ✦
ORTHOPHOSPHORIC ACID ✦ PHOSPHORSAEURELOESUNGEN (GERMAN)

PHOSPHOROUS

Products and Uses: Red: safety matches. White: a rodenticide (rat and mouse poison). In agricultural fertilizers.

Precautions: Red: a relatively nontoxic form. White: a highly toxic form. When swallowed small amounts may cause bloody diarrhea, severe stomach and intestine irritation, liver damage, circulatory collapse, coma, convulsions, and death.

Synonyms: CAS: 7723-14-0 ✦ PHOSPHOROUS, AMORPHOUS

PHOSPHOROUS, OXYCHLORIDE

Products and Uses: A flame-retarding agent for textiles and canvas materials. Used for fireproofing.

Precautions: Toxic by breathing and swallowing. A strong irritant to skin and tissue.

Synonyms: CAS: 10025-87-3 ✦ PHOSPHORYL CHLORIDE

PHYTOCHEMICALS

Products and Uses: A group of substances that occur naturally in fruits, grains, and vegetables. They are not required for body functioning as vitamins and minerals are. Recent research has shown them to be beneficial to the body as cancer and heart disease preventatives. Some that have been in use for years are quinine and digitalis. More recent discoveries include lycopene from tomatoes and the isoflavones from soybeans.

Precautions: Considered to be safe and healthful.

Synonyms: NONE FOUND.

PINE NEEDLE OIL

Products and Uses: An additive in cleaning materials, deodorizers, air fresheners, soaps, toiletries.

Precautions: Could cause allergic reaction. Mildly toxic by swallowing. A human skin irritant. FDA permits use at moderate levels to accomplish the intended effect.

Synonyms: CAS: 8000-26-8 ✦ DWARF PINE NEEDLE OIL ✦ KNEE PINE OIL ✦ LATSCHENKIEFEROL ✦ OIL OF MOUNTAIN PINE ✦ PINUS MONTANA OIL ✦ PINUS PUMILIO OIL ✦ YARMOR

PINE TAR

Products and Uses: Derived from softwood pine trees. Products include tar soap, shampoos, bath oils, and cough medications. Used as a preservative, disinfectant, or deodorant.

Precautions: Can cause allergic reaction.

Synonyms: *PINUS PALUSTRIS* ✦ *PINACEAE*

PIPERONAL

Products and Uses: A chemical used in perfumes, suntan preparations, lipstick, mosquito repellent, and pediculicide (kills lice). In other concentrations it is used as a floral flavoring agent in food products.

Precautions: Moderately toxic by swallowing. Can cause central nervous system depression. A skin irritant. FDA approves use within limitations.

Synonyms: CAS: 120-57-0 ✦ 3,4-BENZODIOXOLE-5-CARBOXALDEHYDE ✦ 3,4-DIHYDROXYBENZALDEHYDE METHYLENE KETAL ✦ DIOXYMETHYLENE-PROTOCATECHUIC ALDEHYDE ✦ HELIOTROPIN ✦ 3,4-METHYLENE-DIHYDROXYBENZALDEHYDE ✦ 3,4-METHYLENEDIOXYBENZALDEHYDE ✦ PIPERONALDEHYDE ✦ PIPERONYL ALDEHYDE ✦ PROTOCATECHUIC ALDEHYDE METHYLENE ETHER

PLASTER OF PARIS

Products and Uses: In baked goods, canned potatoes, canned tomatoes, carrots (canned), confections, frostings, frozen dairy dessert mixes, frozen dairy desserts, gelatins, grain products, ice cream (soft serve), lima beans (canned), pasta, peppers (canned), puddings, wine (sherry), paints (pigment, filler), surgical casts, gypsum board (drywall), and quick-setting cements. It is an anticaking agent, color, coloring agent, dietary supplement, dough conditioner, dough strengthener, drying agent, firming agent, flour-treating agent, leavening agent, nutrient supplement, sequestrant, stabilizer, texturizing agent, thickening agent, and yeast food.

Precautions: Harmless when used for intended purposes.

Synonyms: CAS: 7778-18-9/10101-41-4 ✦ GYPSUM ✦ CALCIUM SULFATE

POLYBUTYLENE

Products and Uses: Found in hot-melt adhesives, packaging tapes, special sealants, and coatings. Used as a gluing, sealing, sticking, adhering material.

Precautions: Items should only be used for intended purposes while closely adhering to label directions.

Synonyms: POLYBUTENE ✦ POLYISOBUTYLENE ✦ POLYISBUTENE

POLYDEXTROSE

Products and Uses: An additive in baked goods, baking mixes (fruit, custard, and pudding-filled pies), cakes, candy (hard), candy (soft), chewing gum, confections, cookies, fillings, frostings, frozen dairy desserts, frozen dairy dessert mixes, gelatins, puddings, and salad dressings. Used as a bulking agent (a filler), formulation aid, humectant (moisturizer), and texturizing agent.

Precautions: FDA requires special labeling for single servings containing above 15 g; "Sensitive individuals may experience a laxative effect from excessive consumption of this product."

Synonyms: CAS: 68424-04-4

POLYETHYLENE

Products and Uses: A protective coating to maintain freshness and appearance on avocados, bananas, beets, Brazil nuts, chestnuts, chewing gum, coconuts, eggplant, filberts, garlic, grapefruit, hazelnuts, lemons, limes, mangoes, muskmelons, onions, oranges, papaya, peas (in pods), pecans, pineapples, plantain, pumpkin, rutabaga, squash (acorn), sweet potatoes, tangerines, turnips, walnuts, and watermelon. Used in hand lotions and as a masticatory (chewing) substance. It is also useful for fabric finishing.

Precautions: FDA approves use for intended purposes.

Synonyms: CAS: 9002-88-4 ✦ AGILENE ✦ ALKATHENE ✦ BAKELITE DYNH ✦ DIOTHENE ✦ ETHENE POLYMER ✦ ETHYLENE HOMOPOLYMER ✦ ETHYLENE POLYMERS ✦ HOECHST PA 190 ✦ MICROTHENE ✦ POLYETHYLENE AS ✦ POLYWAX 1000 ✦ TENITE 800

POLYETHYLENE GLYCOL

Products and Uses: Frequently used in beverages (carbonated), citrus fruit (fresh), sodium nitrite coating, sweeteners (nonnutritive), tablets, and vitamin or mineral preparations. It is a binding agent, coating agent, dispersing agent, flavoring additive, lubricant, and plasticizing agent. Also a softener and humectant in ointments, polish, and base for cosmetics and pharmaceuticals.

Precautions: Slightly toxic by swallowing. An eye irritant. FDA approves when used within limitations.

Synonyms: CAS: 25322-68-3 ✦ CARBOWAX ✦ α-HYDROXY-ω-HYDROXY-POLY(OXY-1,2-ETHANEDIYL) ✦ JEFFOX ✦ LUTROL ✦ PEG ✦ POLY(ETHYLENE OXIDE) ✦ POLY-G SERIES ✦ POLYOX

POLYETHYLENE TEREPHTHALATE

Products and Uses: A filament blended with cotton to produce wash-and-wear fabrics. It is blended with wool to produce worsteds and suitings. Used for recording tapes, soft drink bottles and for surgical grafting material.

Precautions: Harmless when used for intended purposes.

Synonyms: CAS: 25038-59-9 ✦ ALATHON ✦ AMILAR ✦ CELANAR ✦ ESTAR ✦ DACRON ✦ FORTREL ✦ LAVSAN ✦ PEGOTERATE ✦ MYLAR ✦ TERFAN ✦ TERGAL ✦ MELIFORM

POLYSTYRENE COPOLYMER EMULSIONS

Products and Uses: Used for coating and cleaning materials, for example in floor polishes, adhesives, leather protectors, shoe polishes, and cosmetics.

Precautions: An eye, nose, and throat irritant.

Synonyms: URETHANE PREPOLYMERS

POLYURETHANE

Products and Uses: Material used in sealants, caulking, mortars, and adhesives. The consumer products come in both rigid and flexible form. The variety of products range from spandex fibers to cigarette filters to marine flotation devices.

Precautions: Produces toxic fumes upon ignition.

Synonyms: CAS: 9009-54-5 ✦ POLYFOAM PLASTIC ✦ POLYFOAM SPONGE ✦ POLYURETHANE ESTER FOAM ✦ POLYURETHANE SPONGE

POLYVINYL ALCOHOL

Products and Uses: An emulsifier (stabilizes and maintains mixes), thickener, coating, binder, and stabilizer (used to keep a uniform consistency). Useful in cosmetics, cements, mortars, inks, and greaseproofing.

Precautions: Harmless when used for intended purposes.

Synonyms: CAS: 9002-89-5 ✦ PVA ✦ PVOH ✦ ELVANOL ✦ ETHENOL HOMOPOLYMER ✦ VINYL ALCOHOL POLYMER

POLYVINYLPYRROLIDNONE HOMOPOLYMER

Products and Uses: An ingredient in beer, citrus fruit (fresh), confectionery, flavor concentrates in tablet form, food supplements in tablet form, fruits, gum, sweeteners (nonnutritive in concentrated liquid form), sweeteners (nonnutritive in tablet form), vegetables, vinegar, vitamin and mineral concentrates in liquid form, vitamin and mineral concentrates in tablet form, and wine. Also added in adhesives, cosmetics, shampoos, hand creams, skin lotions, dentifrices, and hair sprays. Also useful as a bodying agent (gives body to product), clarifying agent, color fixative, dispersing agent (used to spread ingredients through product), stabilizer (used to keep a uniform consistency), and tableting aid (holds ingredients together).

Precautions: FDA approves food use within limitations.

Synonyms: CAS: 9003-39-8 ✦ ALBIGEN A ✦ ALDACOL Q ✦ BOLINAN ✦ 1-ETHENYL-2-PYRROLIDINONE HOMOPOLYMER ✦ 1-ETHENYL-2-PYRROLIDINONE POLYMERS ✦ HEMODESIS ✦ HEMODEZ ✦ KOLLIDON ✦ LUVISKOL ✦ NEOCOMPENSAN ✦ PERAGAL ST ✦ PERISTON ✦ PLASDONE ✦ POLYCLAR L ✦ POLY(1-(2-OXO-1-PYRROLIDINYL)ETHYLENE) ✦ POLYVIDONE ✦ POLY(n-VINYLBUTYROLACTAM) ✦ PROTAGENT ✦ PVP ✦ SUBTOSAN ✦ VINISIL ✦ N-VINYLBUTYROLACTAM POLYMER ✦ N-VINYLPYRROLIDONE POLYMER

POTASSIUM

Products and Uses: A heart healthy nutrient found naturally in milk, bananas, oranges, potatoes, tomatoes, and other vegetables and fruits. Recent studies indicate that people who took at least 2.5 g potassium supplement reduced their blood pressure, thus reducing the heart's effort in working.

Precautions: Adhere to recommended allowances as listed on container.

Synonyms: K

POTASSIUM ACID TARTRATE

Products and Uses: An ingredient in baked goods, candy (hard), candy (soft), confections, crackers, frostings, gelatins, jams, jellies, margarine, puddings, and wine (grape). It is an acid, anticaking agent, antimicrobial agent (destroys germs), humectant (maintains moisture content), leavening agent (aids in mixing ingredients), processing aid, stabilizer (used to maintain uniform consistency), and thickening agent.

Precautions: FDA permits use within limitations.

Synonyms: CAS: 868-14-4 ✦ CREAM of TARTER ✦ POTASSIUM BITARTRATE

POTASSIUM ARSENATE

Products and Uses: Uses include fly paper, insecticidal preparations, leather tanning, and printing textiles.

Precautions: Very toxic as are all arsenic compounds. Confirmed human carcinogen (causes cancer). Can cause a variety of skin problems including itching, skin pigmentation changes, and skin cancers. A possible mutagen (changes inherited characteristics).

Synonyms: CAS: 7784-41-0 ✦ MACQUER'S SALT ✦ MONOPOTASSIUM ARSENATE ✦ POTASSIUM ACID ARSENATE ✦ POTASSIUM DIHYDROGEN ARSENATE ✦ MONOPOTASSIUM DIHYDROGEN ARSENATE

POTASSIUM BENZOATE

Products and Uses: A preservative added to margarine, and wine.

Precautions: Harmless when used for intended purposes. USDA approves use within limitations.

Synonyms: CAS: 582-25-2

POTASSIUM BROMATE

Products and Uses: An additive in baked goods, beverages (fermented malt), confectionery products, permanent wave solutions, toothpaste, and mouth wash. Used as a dough conditioner, maturing agent (speeds up the aging process in order to make more manageable dough); as an antiseptic and astringent.

Precautions: A poison by swallowing in concentrated amounts. An irritant to skin, eyes, nose, and throat. FDA approves use within limitations.

Synonyms: CAS: 7758-01-2 ✦ BROMIC ACID, POTASSIUM SALT

POTASSIUM CHLORIDE

Products and Uses: Frequently used as a dietary supplement, flavor enhancer, flavoring agent, gelling agent, nutrient, salt substitute, and tissue softening agent, in fertilizer, jelly (artificially sweetened), meat (raw cuts), poultry (raw cuts), preserves (artificially sweetened), and yeast food.

Precautions: FDA states GRAS (generally recognized as safe) when used within specific limitations. However, in pure form it is a human poison by swallowing. Effects on the human body by swallowing are nausea, blood clotting changes, and cardiac arrhythmias. An eye irritant. A mutagen (changes inherited characteristics).

Synonyms: CAS: 7447-40-7 ✦ CHLOROPOTASSURIL ✦ DIPOTASSIUM DICHLORIDE ✦ EMPLETS POTASSIUM CHLORIDE ✦ ENSEAL ✦ KALITABS ✦ KAOCHLOR ✦ KAON-Cl ✦ KAY CIEL ✦ K-LOR ✦ KLOTRIX ✦ K-PRENDE-DOME ✦ PFIKLOR ✦ POTASSIUM MONOCHLORIDE ✦ POTAVESCENT ✦ REKAWAN ✦ SLOW-K ✦ TRIPOTASSIUM TRICHLORIDE

POTASSIUM HYDROXIDE

Products and Uses: A chemical added in soap, bleach, drain cleaners, liquid fertilizers, oven cleaners, paint remover, varnish removers, cosmetic cuticle remover, shaving lotions, hand creams, and facial blushes.

Precautions: Toxic by swallowing and breathing. A strong caustic that is corrosive to tissue in concentrated form. Consumer products contain dilute concentrations.

Synonyms: CAS: 1310-58-3 ✦ CAUSTIC POTASH ✦ POTASSIUM HYDRATE ✦ LYE

POTASSIUM LACTATE

Products and Uses: Used on meat and poultry products, except infant foods and infant formulas. It is a flavor enhancer, flavoring agent, and moisturizing agent.

Precautions: Harmless when used for intended purposes and within approved limits.

Synonyms: CAS: 996-31-6

POTASSIUM LAURATE

Products and Uses: An ingredient in liquid soaps and shampoos. It is an emulsifier (stabilizes and maintains mixes), and a common base.

Precautions: Harmless when used for intended purposes.

Synonyms: NONE FOUND.

POTASSIUM PERMANGANATE

Products and Uses: Frequently used in topical antibacterials, deodorizers, bleaches, and water purification. Used in the production of drugs of abuse, disinfectants, and to bleach stone-washed blue jeans.

Precautions: A human poison by swallowing. Effects on the human body by swallowing are dyspnea (shortness of breath), nausea and gastrointestinal (stomach and intestines) effects. A strong irritant.

Synonyms: CAS: 7722-64-7 ✦ CAIROX ✦ CHAMELEON MINERAL ✦ CONDY'S CRYSTALS

POTASSIUM PEROXYMONOSULFATE

Products and Uses: Utilized in dry laundry bleaches, detergents, washing compounds, cleansers, scouring powders, plastic dishware cleaner, metal cleaners, permanent wave neutralizers, and antiseptics.

Precautions: Harmless when used for intended purposes while adhering to label directions.

Synonyms: OXONE

POTASSIUM SORBATE

Products and Uses: Usually used in baked goods, beverages (carbonated), beverages (still), bread, cake batters, cake fillings, cake topping, cheese, cottage cheese (creamed), smoked fish, salted fish, fresh fruit juices, dried fruits, margarine, pickled goods, pie crusts, pie fillings, salad dressings, salads (fresh), sausage (dry), seafood cocktail, syrups (chocolate dairy), and wine. As a mold retardant, bacteriostat (kills germs), and preservative.

Precautions: Mildly toxic by swallowing. A possible mutagen (changes inherited characteristics). FDA states GRAS (generally recognized as safe) when used within limitations.

Synonyms: CAS: 590-00-1 ✦ 2,4-HEXADIENOIC ACID POTASSIUM SALT ✦ SORBIC ACID, POTASSIUM SALT ✦ SORBISTAT-K ✦ SORBISTAT-POTASSIUM

POTASSIUM STEARATE

Products and Uses: A chemical used in fabric softener, chewing gum, and packaging materials. It is an anticaking agent, binder, emulsifier (stabilizes and maintains mixes), and stabilizer (used to keep a uniform consistency).

Precautions: Can cause allergic reaction in susceptible individuals. FDA approves when used within limitations.

Synonyms: CAS: 593-29-3 ✦ STEARIC ACID POTASSIUM SALT

POTASSIUM SULFITE

Products and Uses: An additive in food and wine that is a preservative.

Precautions: GRAS (generally recognized as safe).

Synonyms: CAS: 10117-38-1 ✦ SULFUROUS ACID, DIPOTASSIUM SALT

POTASSIUM TETROXALATE

Products and Uses: A cleaner and polish in spot remover, rust remover, ink remover, and metal polishes.

Precautions: Products are harmless when used according to label directions.

Synonyms: CAS: 127-96-8 ✦ POTASSIUM QUADROXALATE ✦ SAL ACETOSELLA ✦ SALT OF SORREL

POTASSIUM UNDECYLENATE

Products and Uses: An ingredient added in cosmetics, skin lotions, facial makeup, mascara, eye shadow, and eyeliner pencils. Used in medicinal pharmaceutical items such as skin ointments, creams, and lotions as a bacteriostat (kills bacteria) and fungistat (kills fungus).

Precautions: Toxic in high concentrations.

Synonyms: NONE FOUND.

PROPANE

Products and Uses: Commonly used as an aerating agent, gas refrigerant, and spray propellant for cosmetics, shaving creams, foamed foods, and sprayed foods.

Precautions: Central nervous system effects at high concentrations. An asphyxiant. Flammable gas. GRAS (generally recognized as safe) by FDA.

Synonyms: CAS: 74-98-6 ✦ DIMETHYLMETHANE ✦ PROPYL HYDRIDE

PROPIONIC ACID

Products and Uses: An ingredient used in perfumes, artificial fruit flavors, breads, and grains. Antimicrobial agent, flavoring agent, mold inhibitor in bread, preservative, and emulsifying agent (stabilizes and maintains mixes).

Precautions: Moderately toxic by swallowing and skin contact. A corrosive irritant to eyes, skin, nose, and throat. FDA states GRAS (generally recognized as safe)

Synonyms: CAS: 79-09-4 ✦ CARBOXYETHANE ✦ ETHANECARBOXYLIC ACID ✦ ETHYLFORMIC ACID ✦ METACETONIC ACID ✦ METHYL ACETIC ACID ✦ PROPANOIC ACID ✦ PROPIONIC ACID, solution containing not less than 80% acid ✦ PROPIONIC ACID GRAIN PRESERVER ✦ PROZOIN ✦ PSEUDOACETIC ACID ✦ SENTRY GRAIN PRESERVER ✦ TENOX P GRAIN PRESERVATIVE

PROPYLENE GLYCOL MONO- and DIESTERS

Products and Uses: An aerating agent, gas refrigerant, and spray propellant. For beet sugar, cake batters, cake icings, cake shortening, margarine, oil, poultry fat (rendered), whipped toppings, and yeast. As an emulsifier (stabilizes and maintains mixes) and stabilizer (used to keep a uniform consistency).

Precautions: FDA and USDA permit use within limitations.

Synonyms: PROPYLENE GLYCOL MONO- and DIESTERS OF FATTY ACIDS ✦ PROPYLENE GLYCOL MONOSTEARATE

PROPYLENE GLYCOL PHENYL ETHER

Products and Uses: A bactericide and fixative for fragrance and cleaning products in soaps and perfumes.

Precautions: Harmless when used for intended purposes.

Synonyms: NONE FOUND.

PROPYLPARABEN

Products and Uses: Commonly used as a preservative, fungicide, and mold preventer. It is an antimicrobial (kills germs) in food and pharmaceutical products.

Precautions: Approved by FDA. Mildly toxic by swallowing. An allergen. Could cause allergic reaction.

Synonyms: CAS: 94-13-3 ✦ PROPYLPARASEPT ✦ PARABEN ✦ PARASEPT ✦ ASEPTOFORM ✦ BETACIDE ✦ PRESERVAL ✦ p-HYDROXYPROPYL BENZOATE

PUMICE

Products and Uses: Usually in cleansers, scouring agents, fireproofing and insulating compounds, cosmetics for removing rough skin, heavy-duty hand soaps, facial cleansing and acne compounds, tooth polishes, and denture powders. Pencil erasers are composed of synthetic rubber and pumice. (It is the pumice that erases, not the rubber.) Various abrasive purposes.

Precautions: Harmless when used for intended purposes.

Synonyms: SEISMOTITE

PYRETHRIN

Products and Uses: An aerating agent, gas refrigerant, and spray propellant. Added to animal feed, dried foods, milled fractions derived from cereal grains, packaging materials. Also utilized in insecticides for treating food shipping materials. It is a household pesticide.

Precautions: Moderately toxic by swallowing. Can cause allergic reactions. May cause nausea, vomiting, headache and other effects. FDA permits use within limitations.

Synonyms: CAS: 97-11-0 ✦ 2-CYCLOPENTENYL-4-HYDROXY-3-METHYL-2-CYCLOPENTEN-1-ONE CHRYSANTHEMATE ✦ 3-(2-CYCLOPENTEN-1-YL)-2-METHYL-4-OXO-2-CYCLOPENTEN-1-YL CHRYSANTHEMUMATE ✦ 3-(2-CYCLOPENTENYL)-2-METHYL-4-OXO-2-CYCLOPENTENYL CHRYSANTHEMUMMONOCARBOXYLATE ✦ CYCLOPENTENYLRETHONYL CHRYSANTHEMATE

QUANTERNARY AMMONIUM SALT

Products and Uses: Utilized as an aerating agent, gas refrigerant, and spray propellant in cosmetics, after-shave lotions, deodorants, hair colorings, hair curling products, dandruff removers, and detergents. Used as a disinfectant, cleanser, sterilizer, for fungicide and mildew control, and others.

Precautions: A poison by swallowing. A skin and severe eye irritant.

Synonyms: CAS: 8001-54-5 ✦ BENZALKONIUM CHLORIDE ✦ DRAPOLENE ✦ PHENEENE GERMICIDAL SOLUTION ✦ ZEPHIRAN CHLORIDE

QUERCETIN

Products and Uses: An antioxidant chemical found in onions and apples.

Precautions: This is considered a beneficial chemical.

Synonyms: NONE KNOWN.

QUERTINE

Products and Uses: An additive in hair products, used as a dye or coloring agent.

Precautions: Poison by swallowing. A possible carcinogen (causes cancer). A human mutagen (changes inherited characteristics). Could cause allergic reactions in susceptible individuals.

Synonyms: CAS: 117-39-5 ✦ C.I. NATURAL RED ✦ C.I. NATURAL YELLOW ✦ MELETIN ✦ QUERCETOL ✦ QUERCITINE ✦ SOPHORETIN ✦ XANTHAURINE ✦ CYANDIELONON

QUILLAJA EXTRACT

Products and Uses: Various applications include in shampoo, bubble baths, soap products, detergents, soap substitutes; in mineral water, root beer, beverages, spices, candies, and dessert ices. Also used as a foam producer in cleaning products and in fire extinguishers. It is also a flavoring ingredient.

Precautions: Harmless when used for intended purposes.

Synonyms: SOAP BARK ✦ QUILLAY BARK ✦ PANAMA BARK ✦ CHINA BARK ✦ MURILLO BARK

QUININE

Products and Uses: A substance included in tonics, medications (antimalarial), beverages, and cosmetics. Used as a pharmaceutical and flavoring agent.

Precautions: Effects on the human body by swallowing are visual changes, tinnitus (ringing in the ears), nausea, or vomiting. In concentration it can cause mutagenic (changes inherited characteristics), and teratogenic (abnormal fetal development) effects. A skin irritant, particularly to barbers and beauticians who work with quinine tonics.

Synonyms: CAS: 130-95-0 ✦ 6-METHOXYCINCHONINE

QUINOLINE YELLOW

Products and Uses: A colorant ingredient used in drugs, cosmetics, facial makeup, and skin creams.

Precautions: Synthetic dye approved for use by FDA except in eye area.

Synonyms: C.I. ACID YELLOW ✦ D&C YELLOW NO. 10 ✦ ACID YELLOW 3 ✦ FOOD YELLOW 13

RADON

Products and Uses: The radioactive gas of radium and uranium decay. Used in medical treatment and chemical research.

Precautions: A common air contaminant. Accumulation of the gas in homes is suspected to be the second leading cause of lung cancer in the U.S. This accumulation is found in well-insulated buildings located over land which has concentrations of uranium. Ventilation prevents accumulation.

Synonyms: CAS: 10043-92-2 ✦ Rn

RENNET

Products and Uses: Frequently in fillings, frozen dairy desserts, gelatins, loaves (nonspecific), milk products, poultry, puddings, sausage, sausage (imitation), soups, and stews. As a binder, enzyme, extender, processing aid, stabilizer (keeps a uniform consistency), and thickening agent.

Precautions: FDA states GRAS (generally recognized as safe).

Synonyms: CAS: 9001-98-3 ✦ BOVINE RENNET

RESORCINOL

Products and Uses: A chemical in adhesives, cosmetics, hair coloring, antidandruff shampoos, lipstick, and hair care products. Used as an antiseptic, antipruritic (anti-itching agent), preservative, dye, and astringent.

Precautions: A possible allergen. Could be an irritant to eyes, nose, throat, and skin.

Synonyms: CAS: 108-46-3 ✦ RESORCIN ✦ m-DIHYDROXYBENZENE ✦ 3-HYDROXYPHENOL

RESVERATROL

Products and Uses: An ingredient in red wines and grape juice. It is believed to be the ingredient in grapes that lowers artery-clogging low-density lipoprotein (LDL) cholesterol, while elevating good high-density lipoprotein (HDL) cholesterol. The HDL helps flush coronary arteries of fatty deposits.

Precautions: Harmless when used in moderate levels.

Synonyms: NONE FOUND.

RHODINOL

Products and Uses: An ingredient in food, beverages, and cosmetics. Produces a synthetic roselike aroma. An ingredient to affect the taste or smell of products.

Precautions: Could cause allergic reaction in susceptible individuals.

Synonyms: CAS: 6812-78-8 ✦ *l*-CITRONELLOL ✦ REUNION GERANIUM OIL

RIBOFLAVINE

Products and Uses: An ingredient added to peanut butter, cereals, bread stuffing, flour, grits, cornmeal, pastas, and bread products. Occurs naturally in milk, eggs, and organ meats. Used as a color additive, dietary supplement, and nutritional supplement.

Precautions: FDA states GRAS (generally recognized as safe).

Synonyms: CAS: 83-88-5 ✦ BEFLAVINE ✦ 6,7-DIMETHYL-9-*d*-RIBITYLISOALLOXAZINE ✦ 7,8-DIMETHYL-10-*d*-RIBITYLISOALLOXAZINE ✦ 7,8-DIMETHYL-10-(*d*-RIBO-2,3,4,5-TETRAHYDROXYPENTYL)ISOALLOXAZINE ✦ FLAVAXIN ✦ HYFLAVIN ✦ HYRE ✦ LACTOFLAVIN ✦ LACTOFLAVINE ✦ RIBIPCA ✦ RIBODERM ✦ RIBOFLAVIN ✦ RIBOFLAVINEQUINONE ✦ VITAMIN B2 ✦ VITAMIN G

RICE BRAN WAX

Products and Uses: Utilized in candy, chewing gum, fruits (fresh), and vegetables (fresh). It is a coating agent, masticatory (chewing) substance, and release agent.

Precautions: FDA approves use within limitations.

Synonyms: NONE FOUND.

RICINOLEIC ACID

Products and Uses: An additive in soaps, dry-cleaning soaps, castor oil, fabric sizing, and in skin softener lotions. Used for fabric conditioning, cleaning, and finishing. It is also a skin moisturizer.

Precautions: A possible carcinogen (cancer-causing agent).

Synonyms: CAS: 141-22-0 ✦ RICINOLIC ACID ✦ RICINIC ACID ✦ CASTOR OIL ACID ✦ 12-HYDROXYOLEIC ACID

ROSIN

Products and Uses: Derived from pine trees. Used in adhesives, varnishes, soaps, inks, rosin bags, and mastics.

Precautions: Harmless when used for intended purposes.

Synonyms: GUM ROSIN ✦ WOOD ROSIN ✦ TALL OIL ROSIN

RUE OIL and HERB

Products and Uses: A seasoning used as a fruit and spice flavoring in baked goods, candy (soft), frozen dairy desserts, and mixes. An ingredient that affects taste and smell.

Precautions: A possible allergen. FDA states GRAS (generally recognized as safe) when used within limitations.

Synonyms: NONE FOUND.

SACCHARIN

Products and Uses: A nonnutritive synthetic sweetener in bacon, bakery products, beverage mixes, beverages, chewing gum, desserts, fruit juice drinks, jam, relishes, breath fresheners, and chewable vitamin tablets.

Precautions: Slightly toxic by swallowing. A human mutagen (changes inherited characteristics). FDA permits use within limitations. The National Academy of Sciences has stated that saccharin is a potential carcinogen (causes cancer) and products containing it must have a warning label because it caused cancer in lab animals. Pregnant women should avoid heavy use of this sweetener.

Synonyms: CAS: 128-44-9 ✦ CRYSTALLOSE ✦ SACCHARINE, SODIUM ✦ SAXIN ✦ SODIUM BENZOSULPHIMIDE ✦ SOLUBLE GLUSIDE ✦ SUCCARIL ✦ SUCRA ✦ SYKOSE

SAFFLOWER OIL

Products and Uses: Various uses include coating and emulsifying agent (stabilizes and maintains mixes), formulation aid, and texturizing agent in resins, paints, varnishes, dietetic foods, margarine, and hydrogenated shortening.

Precautions: A possible skin irritant. Swallowing of large doses can cause vomiting. GRAS (generally recognized as safe).

Synonyms: CAS: 8001-23-8 ✦ SAFFLOWER OIL (UNHYDROGENATED)

SAFROL

Products and Uses: Useful as a fragrance (sassafras or camphorwood aroma) in cosmetics, soaps, and perfumes.

Precautions: A confirmed carcinogen (causes cancer). Moderately toxic by swallowing. A mutagen (changes inherited characteristics). A skin irritant. FDA prohibits use in food.

Synonyms: CAS: 94-59-7 ✦ 5-ALLYL-1,3-BENZODIOXOLE ✦ ALLYLCATECHOL METHYLENE ETHER ✦ ALLYLDIOXYBENZENE METHYLENE ETHER ✦ 1-ALLYL-3,4-METHYLENEDIOXYBENZENE ✦ 4-ALLYL-1,2-METHYLENEDIOXYBENZENE ✦ m-ALLYLPYROCATECHIN METHYLENE ETHER ✦ 4-ALLYLPYROCATECHOL FORMALDEHYDE ACETAL ✦ ALLYLPYROCATECHOL METHYLENE ETHER ✦ 1,2-METHYLENEDIOXY-4-ALLYLBENZENE ✦ 3,4-METHYLENEDIOXY-ALLYBENZENE ✦ 5-(2-PROPENYL)-1,3-BENZODIOXOLE ✦ RHYUNO OIL ✦ SAFROLE ✦ SAFROLE MF ✦ SHIKIMOLE ✦ SHIKOMOL

SALICYLIC ACID

Products and Uses: A chemical in aspirin and over-the-counter drugs, used as an analgesic (pain-reliever).

Precautions: Poison by swallowing. Effects on the human body by skin contact are ear tinnitus (ringing). A mutagen (changes inherited characteristics). A skin and severe eye irritant.

Synonyms: CAS: 69-72-7 ✦ o-HYDROXYBENZOIC ACID ✦ 2-HYDROXYBENZOIC ACID ✦ KERALYT ✦ ORTHOHYDROXYBENZOIC ACID ✦ RETARDER W ✦ SA

SAPONIN

Products and Uses: A component in soaps, shampoos, detergents, shaving creams, fire extinguishers, detergents, and bath products as a foam producer and an emulsifier.

Precautions: Very low toxicity when swallowed.

Synonyms: SAPONARIA ✦ QUILLAJA

SASSAFRAS

Products and Uses: A seasoning and scenting agent in bakery products, beverages (nonalcoholic), confections, gelatin desserts, puddings, toothpaste, soaps, perfumes, and powders.

Precautions: A skin irritant. FDA approves use at moderate levels to accomplish the intended effect.

Synonyms: SASSAFRAS ALBIDUM

SELENIUM

Products and Uses: A trace mineral that occurs naturally in whole wheat bread and pastas, Brazil nuts, pork, lamb, and beef. The recommended daily allowance is 55 mcgs. A 1996 study showed that selenium reduced some cancer incidence by 67%. Unfortunately, this was an all-male study. Research still needs to be done in women, on hormone-related cancers.

Precautions: Poisonous if excessive amounts are consumed, usually in the form of supplements.

Synonyms: NONE KNOWN.

SHELLAC

Products and Uses: Derived from the (almost continuous) excrement of insects that feed on resiniferous trees in India. Found in varnishes, self-polishing waxes, abrasives, hair sprays, sealing wax, cements, sealers, cake glazes, confectioneries, food supplements in tablet form, and gum. Used as a coating agent, color fixative, food and candy glaze, and surface-finishing agent for floors and furniture.

Precautions: Could cause allergic reaction in susceptible individuals.

Synonyms: WHITE SHELLAC ✦ REGULAR BLEACHED SHELLAC ✦ WAX-FREE SHELLAC ✦ REFINED BLEACH SHELLAC ✦ LAC ✦ GARNET LAC ✦ GUM LAC ✦ STICK LAC

SILICA

Products and Uses: Multiple applications include in bacon (cured), baking powder, beer, coffee whiteners, egg yolk (dried), flour, fruits, gelatin desserts, pudding mixes, salt, soups (powdered), tortilla chips, vanilla powder, and vegetables. In cosmetics, pharmaceuticals, and food. It is an anticaking agent, antifoaming agent, carrier, malt beverage chillproofing agent, color fixative, conditioning agent, defoaming agent, ink (food marking), and wax to prevent slipping.

Precautions: Moderately toxic by swallowing. Much less toxic than crystalline forms. Does not cause silicosis. FDA permits use within limitations.

Synonyms: CAS: 112945-52-5 ✦ ACTICEL ✦ AEROSIL ✦ AMORPHOUS SILICA DUST ✦ AQUAFIL ✦ CAB-O-GRIP II ✦ CAB-O-SIL ✦ CAB-O-SPERSE ✦ CATALOID ✦ COLLOIDAL SILICA ✦ COLLOIDAL SILICON DIOXIDE ✦ DICALITE ✦ DRI-DIE INSECTICIDE 67 ✦ FLO-GARD ✦ FOSSIL FLOUR ✦ FUMED SILICA ✦ FUMED SILICON DIOXIDE ✦ HI-SEL ✦ LO-VEL ✦ LUDOX ✦ NALCOAG ✦ NYACOL ✦ SANTOCEL ✦ SILICA AEROGEL ✦ SILICA, AMORPHOUS ✦ SILICIC ANHYDRIDE ✦ SILICON DIOXIDE ✦ SILIKILL ✦ SYNTHETIC AMORPHOUS SILICA ✦ VULKASIL

SODIUM ALUMINOSILICATE

Products and Uses: Added to cake mixes, mixes (dry), nondairy creamers, salt, and sugar (powdered). Useful as an anticaking agent.

Precautions: FDA states GRAS (generally recognized as safe) within limitations. An irritant to skin, eyes, nose, and throat.

Synonyms: CAS: 1344-00-9 ✦ SODIUM SILICOALUMINATE

SODIUM BENZOATE

Products and Uses: It is an antimicrobial (kills germs) agent; preservative for food and pharmaceuticals. Useful as an additive in tobacco, creams, carbonated drinks, pickles, toothpaste, fruit juice, preserves, margarine.

Precautions: FDA states GRAS (generally recognized as safe) when used within limitations. Moderately toxic by swallowing. Small doses have little or no effect.

Synonyms: CAS: 532-32-1 ✦ ANTIMOL ✦ BENZOATE OF SODA ✦ BENZOATE SODIUM ✦ BENZOESAEURE ✦ BENZOIC ACID, SODIUM SALT ✦ SOBENATE ✦ SODIUM BENZOIC ACID

SODIUM BICARBONATE

Products and Uses: A common chemical used in baked goods, beverages (dry mix), fats (rendered), margarine, pickles (cured), soups, poultry, dairy products, and vegetables. An ingredient in mouthwashes, bath products, and skin medication products. It is an alkali, gastric antacid, cleaning agent, and leavening agent.

Precautions: Effects on the human body by swallowing are respiratory changes, increased urine volume, and sodium level changes. A mutagen (changes inherited characteristics). FDA states GRAS (generally recognized as safe) when used in appropriate amounts. USDA approves use sufficient for purpose.

Synonyms: CAS: 144-55-8 ✦ BAKING SODA ✦ BICARBONATE OF SODA ✦ CARBONIC ACID MONOSODIUM SALT ✦ COL-EVAC ✦ JUSONIN ✦ MONOSODIUM CARBONATE ✦ NEUT ✦ SODA MINT ✦ SODIUM ACID CARBONATE ✦ SODIUM HYDROGEN CARBONATE

SODIUM BISULFITE

Products and Uses: A preservative that prevents discoloration of dried fruit, fresh shrimp, dried, fried, frozen potatoes, and prevents bacterial growth in wine.

Precautions: Moderately toxic by swallowing. A corrosive irritant to skin, eyes, nose and throat. A mutagen (changes inherited characteristics). Causes allergic reaction in susceptible individuals. GRAS (generally recognized as safe).

Synonyms: CAS: 7631-90-5 ✦ BISULFITE DE SODIUM (FRENCH) ✦ HYDROGEN SULFITE SODIUM ✦ SODIUM ACID SULFITE ✦ SODIUM BISULFITE(1:1) ✦ SODIUM BISULFITE, SOLID ✦ SODIUM BISULFITE, SOLUTION ✦ SODIUM HYDROGEN SULFITE ✦ SODIUM HYDROGEN SULFITE, SOLID ✦ SODIUM HYDROGEN SULFITE, SOLUTION ✦ SODIUM SULHYDRATE ✦ SULFUROUS ACID, MONOSODIUM SALT

SODIUM CARBOXYMETHYL CELLULOSE

Products and Uses: An ingredient in baked pie fillings, poultry, ice cream, beer, icings, diet foods, and candies. It is used as a binder (to hold substances together), extender, stabilizer (keeps uniform consistency), and thickener (adds body and thickness to texture).

Precautions: Mildly toxic by swallowing. FDA states GRAS (generally recognized as safe) when used within stated limitations.

Synonyms: CAS: 9004-32-4 ✦ CMC ✦ CARMETHOSE ✦ CELLOFAS ✦ CELLUGEL ✦ CELLULOSE GUM ✦ CELLULOSE SODIUM GLYCOLATE ✦ FINE GUM HES ✦ LUCEL (POLYSACCHARIDE)

SODIUM CASEINATE

Products and Uses: Commonly used in bread, cereals, cheese (imitation), coffee whiteners, desserts, egg substitutes, loaves (nonspecific), meat (processed), poultry, sausage (imitation), soups, stews, whipped toppings, whipped toppings (vegetable oil), and wine. It is a milk protein used as a binder (holds substances together), clarifying agent (removes small particles from liquids), emulsifier (stabilizes and maintains mixes), extender, and stabilizer (keeps a uniform consistency).

Precautions: FDA states GRAS (generally recognized as safe). USDA approves use within limitations. BATF (Bureau of Alcohol Tax and Firearms) states GRAS (generally recognized as safe).

Synonyms: CAS: 9005-46-3 ✦ CASEIN and CASEINATE SALTS ✦ CASEIN-SODIUM ✦ CASEIN, SODIUM COMPLEX ✦ CASEINS, SODIUM COMPLEXES ✦ NUTROSE

SODIUM CHLORIDE

Products and Uses: Its many uses include over-the-counter pharmaceuticals, antiseptics, and astringents for skin abrasions. Used in soaps, bath, and dental products. In baked goods, butter, cheese, nuts (salted), poultry, and sausage. In mineral waters and home water softeners. Frequently used as chilling media, curing agent, dough conditioner, flavoring agent, intensifier, nutrient, and preservative and for ice and snow control.

Precautions: Moderately toxic by swallowing. An effect on the human body by swallowing is blood pressure increase. A skin and eye irritant. Swallowing of large amounts of sodium chloride can cause irritation of the stomach. Improper use of salt tablets may produce this effect. USDA states GRAS (generally recognized as safe) when used within limits.

Synonyms: CAS: 7647-14-5 ✦ COMMON SALT ✦ DENDRITIS ✦ EXTRA FINE 200 SALT ✦ EXTRA FINE 325 SALT ✦ HALITE ✦ H.G. BLENDING ✦ NATRIUMCHLORID (GERMAN) ✦ PUREX ✦ ROCK SALT ✦ SALINE ✦ SALT ✦ SEA SALT ✦ STERLING ✦ TABLE SALT ✦ TOP FLAKE ✦ USP SODIUM CHLORIDE ✦ WHITE CRYSTAL

SODIUM CITRATE

Products and Uses: Prevents separation of solids in evaporated milk. Used in soft drinks, frozen desserts, meat products, detergents, and special cheeses. As a buffer (in carbonated beverages); improves whipping properties in cream; an emulsifier (in cheese).

Precautions: Harmless when used for intended purposes.

Synonyms: CAS: 68-04-2 ✦ TRISODIUM CITRATE

SODIUM DIACETATE

Products and Uses: Frequently used as an antimicrobial agent in baked goods, bread, candy (soft), fats, gravies, meat products, oil, sauces, snack foods, soups, as a flavoring agent, mold inhibitor, preservative, and sequestrant, (binds constituents that affect the final product's appearance, flavor or texture).

Precautions: FDA states GRAS (generally recognized as safe) when used within limits.

Synonyms: CAS: 126-96-5 ✦ SODIUM HYDROGEN DIACETATE

SODIUM DICHLOROCYANURATE

Products and Uses: A component in dry bleaches, dishwashing detergents, scouring powder, detergent sanitizers, and pool disinfectants. Used as a replacement for calcium hypochlorite and bleaching materials.

Precautions: Moderately toxic by swallowing. A skin and severe eye irritant. Swallowing causes emaciation, weakness, lethargy, diarrhea, and weight loss.

Synonyms: CAS: 2244-21-5 ✦
SODIUM SALT OF DICHLORO-S-TRIAZINE-2,4,6-TRIONE

SODIUM DODECYL SULFATE

Products and Uses: An ingredient commonly added to carpet shampoos, detergents, and wax products. Included in cleaning products and soil emulsifiers.

Precautions: Moderately toxic by swallowing. A skin irritant. A mild allergen.

Synonyms: CAS: 151-21-3 ✦ AQUAREX METHYL ✦ DODECYL ALCOHOL, HYDROGEN SULFATE, SODIUM SALT ✦ DODECYL SODIUM SULFATE ✦ DODECYLSULFATE SODIUM SALT ✦ DREFT ✦ LAURYL SODIUM SULFATE ✦ SODIUM MONODECYL SULFATE ✦ SULFURIC ACID, MONODECYL ESTER, SODIUM SALT

SODIUM HYDROXIDE

Products and Uses: An active ingredient in drain cleaners, oven cleaners, caustic sodas, lye, black olives, brandy, margarine, meat food products, poultry products, wine spirit (brandies, sherries, and so on). Added to hair processing (straightening) products, cuticle removers, and shaving cream products. It is an emulsifier in liquid cosmetics.

Precautions: Moderately toxic by swallowing. A mutagen (changes inherited characteristics). A corrosive irritant to skin, eyes, nose, and throat. This material, both solid and in solution, has a markedly corrosive action upon all body tissue, causing burns and frequently deep ulceration, with ultimate scarring. Mists, vapors, and dusts of this compound cause small burns, and contact with the eyes rapidly causes severe damage to the delicate tissue. Swallowing causes very serious damage to the nose and throat or other tissues with which contact is made. It can cause perforation and scarring. Breathing of the dust or concentrated mist can cause damage to the upper respiratory tract and to lung tissue, depending upon the severity of the exposure. Thus, effects of breathing may vary from mild irritation of the nose and throat to a severe pneumonitis.

Synonyms: CAS: 1310-73-2 ✦ CAUSTIC SODA ✦ CAUSTIC SODA, BEAD ✦ CAUSTIC SODA, DRY ✦ CAUSTIC SODA, FLAKE ✦ CAUSTIC SODA, GRANULAR ✦ CAUSTIC SODA, LIQUID ✦ CAUSTIC SODA, SOLID ✦ CAUSTIC SODA, SOLUTION ✦ LEWIS-RED DEVIL LYE ✦ LYE ✦ SODA LYE ✦ SODIUM HYDRATE ✦ SODIUM HYDROXIDE, BEAD ✦ SODIUM HYDROXIDE, DRY ✦ SODIUM HYDROXIDE, FLAKE ✦ SODIUM HYDROXIDE, GRANULAR ✦ SODIUM HYDROXIDE, SOLID ✦ WHITE CAUSTIC

SODIUM HYPOCHLORITE

Products and Uses: A commonly used bleach, disinfectant, and drain cleaner. Also used for swimming pool disinfectant, laundry bleaches, water purification, and germicide.

Precautions: A human mutagen (changes inherited characteristics). Corrosive and irritating by swallowing and breathing.

Synonyms: CAS: 7681-52-9 ✦ ANTIFORMIN ✦ CHLOROS ✦ CHLOROX ✦ HYCLORITE ✦ SURCHLOR

SODIUM LAURYL SULFATE

Products and Uses: A controversial chemical used in angel food cake, beverage bases, citrus fruit (fresh), egg white (frozen), egg white (liquid), egg white solids, fruit juice drink, marshmallows, poultry, and vegetable oils. An additive in detergents, shampoos, toothpastes, hand lotions, and bubble baths. An emulsifier (stabilizes and maintains mixes), surfactant, wetting agent, and whipping agent.

Precautions: Moderately toxic by swallowing. A human skin irritant. A mild allergen. A mutagen (changes inherited characteristics).

Synonyms: CAS: 151-21-3 ✦ AQUAREX METHYL ✦ CARSONOL SLS ✦ CONCO SULFATE WA ✦ CYCLORYL 21 ✦ DETERGENT 66 ✦ DODECYL ALCOHOL, HYDROGEN SULFATE, SODIUM SALT ✦ DODECYL SODIUM SULFATE ✦ DODECYL SULFATE, SODIUM SALT ✦ DREFT ✦ DUPONOL ✦ HEXAMOL SLS ✦ IRIUM ✦ LANETTE WAX-S ✦ LAURYL SODIUM SULFATE ✦ LAURYL SULFATE, SODIUM SALT ✦ MAPROFIX 563 ✦ NEUTRAZYME ✦ ORVUS WA PASTE ✦ SIPEX OP ✦ SIPON WD ✦ SLS ✦ SODIUM DODECYL SULFATE ✦ SODIUM MONODODECYL SULFATE ✦ SOLSOL NEEDLES ✦ STERLING WAQ-COSMETIC ✦ SULFOPON WA 1 ✦ SULFOTEX WALA ✦ SULFURIC ACID, MONODODECYL ESTER, SODIUM SALT ✦ ULTRA SULFATE SL-1

SODIUM METAPHOSPHATE

Products and Uses: An ingredient in toothpastes, detergents, and water softeners. Added to cheese, dairy products, fish, lima beans, meat food products, milk, peanuts, peas (canned), and poultry food products. Used as a sequestrant (binds constituents that affect the final product's appearance, flavor or texture), emulsifier (stabilizes and maintains mixes), additive, and for laundering.

Precautions: FDA states GRAS (generally recognized as safe) when used within limits.

Synonyms: CAS: 10361-03-2 ✦ GRAHAM'S SALT ✦ METAFOS ✦ SODIUM HEXAMETAPHOSPHATE ✦ SODIUM POLYPHOSPHATES, GLASSY ✦ SODIUM TETRAPOLYPHOSPHATE

SODIUM NITRITE

Products and Uses: Used for smoked cured tunafish, sablefish, salmon, shad, frankfurters, luncheon meats, bacon, corned beef, ham (canned), meat (cured), and poultry. It is an antimicrobial (kills germs) agent; color fixative in meat and meat products; a preservative.

Precautions: Human poison by swallowing. Effects on the human body by swallowing are motor activity changes, coma, decreased blood pressure with possible pulse rate increase without fall in blood pressure, arteriolar or venous dilation, nausea or vomiting, and blood chemistry changes. Nitrite can lead to the formation of small amounts of cancer-causing chemicals (nitrosamines), particularly in fried bacon. Nitrite is tolerated in foods because it prevents growth of bacteria that cause botulism (poisoning). A possible carcinogen (causes cancer). A human mutagen (changes inherited characteristics). FDA permits use within stated limitations.

Synonyms: CAS: 7632-00-0 ✦ ANTI-RUST ✦ DIAZOTIZING SALTS ✦ ERINITRIT ✦ FILMERINE ✦ NITROUS ACID, SODIUM SALT

SODIUM PALMITATE

Products and Uses: A common ingredient in shaving creams, laundry soaps, toilet soaps, detergents, cosmetics, and inks. It is a texturizer (smooths ingredients) in consumer toiletries and cleaning products. Also useful as an emulsifier (stabilizes and maintains mixes) and stabilizer (keeps uniform consistency). It is an anticaking agent.

Precautions: Harmless when used for intended purposes. FDA states GRAS (generally recognized as safe).

Synonyms: SODIUM SALT OF PALMITIC ACID

SODIUM PERBORATE

Products and Uses: Utilized to sanitize, deodorize, and clean. It is a bleaching agent in special detergents. Used for dentures or partial dental plates.

Precautions: Toxic by swallowing. Strong oxidizing agent (gives off oxygen).

Synonyms: CAS: 7632-04-4 ✦ SODIUM PERBORATE ANHYDROUS ✦ SODIUM PERBORATE MONOHYDRATE

SODIUM PEROXIDE

Products and Uses: Frequently used in various bleaches, deodorants, antiseptics, water purification, dyes, and germicidal soaps. It is for purifying air in sick rooms.

Precautions: A severe irritant to skin, eyes, nose, and throat. Dangerous fire and explosion risk.

Synonyms: CAS: 1313-60-6 ✦ DISODIUM DIOXIDE ✦ DISODIUM PEROXIDE ✦ SODIUM DIOXIDE ✦ SOLOZONE ✦ SODIUM OXIDE

SODIUM PROPIONATE

Products and Uses: A common additive in baked goods, beverages (nonalcoholic), candy (soft), cheese, confections, dough (fresh pie), fillings, frostings, gelatins, jams, jellies, meat products, pizza crust, and puddings. It is an antimicrobial (kills germs) agent, additive, flavoring agent, mold and mildew inhibitor, and preservative.

Precautions: Moderately toxic by skin contact. Mildly toxic by unspecified routes. An allergen. FDA and USDA permits use within limitations.

Synonyms: CAS: 137-40-6 ✦ PROPANOIC ACID, SODIUM SALT

SODIUM SILICATE

Products and Uses: A chemical used to sanitize, deodorize, and clean. It is a mild antiseptic. An additive for eggs, cosmetics, soaps, depilatories, protective creams, topical antiseptics, gels, and adhesives. Utilized as a preservative, anticaking agent, and detergent.

Precautions: Poison by swallowing. A caustic material which is a severe eye, skin, nose, and throat irritant. Swallowing causes gastrointestinal (stomach and intestinal) tract upset. FDA and USDA approves use at moderate levels.

Synonyms: CAS: 6834-92-0 ✦ B-W ✦ CRYSTAMET ✦ DISODIUM METASILICATE ✦ DISODIUM MONOSILICATE ✦ METSO 20 ✦ METSO BEADS 2048 ✦ METSO BEADS, DRYMET ✦ METSO PENTABEAD 20 ✦ ORTHOSIL ✦ SODIUM METASILICATE ✦ SODIUM METASILICATE, ANHYDROUS ✦ WATER GLASS

SORBIC ACID

Products and Uses: Frequently used in baked goods, beverages (carbonated), beverages (still), bread, cake batters, cake fillings, cake topping, cheese, cottage cheese (creamed), fish (smoked or salted), fruit juices (fresh), fruits (dried), margarine, pickled goods, pie crusts, pie fillings, salad dressings, salads (fresh), sausage (dry), seafood cocktail, syrups (chocolate dairy), and wine. Not permitted in cooked sausage and meat salads. It is a preservative and food additive. It prevents mold growth.

Precautions: Mildly toxic by swallowing. A severe human skin irritant. A mutagen (changes inherited characteristics). FDA and USDA approve use within limitations.

Synonyms: CAS: 110-44-1 ✦ (2-BUTENYLIDENE)ACETIC ACID ✦ CROTYLIDENE ACETIC ACID ✦ HEXADIENIC ACID ✦ HEXADIENOIC ACID ✦ 2,4-HEXADIENOIC ACID ✦ trans-trans-2,4-HEXADIENOIC ACID ✦ 1,3-PENTADIENE-1-CARBOXYLIC ACID ✦ 2-PROPENYLACRYLIC ACID ✦ SORBISTAT

SORBITAN MONOSTEARATE

Products and Uses: An additive in cake fillings, cake icings, cake mixes, cakes, chocolate coatings, coffee whiteners, confectionery coatings, cream fillings, desserts (nonstandardized frozen), fruit (raw), milk or cream substitutes for beverage coffee, oil (edible whipped topping), vegetables (raw), and yeast (active dry). Used as a defoaming agent, emulsifier (stabilizes and maintains mixes), rehydration aid (replaces moisture), and stabilizer (helps keep a uniform consistency).

Precautions: Very mildly toxic by swallowing. A skin irritant. FDA approves use within stated limits.

Synonyms: CAS: 1338-41-6 ✦ ANHYDRO-*d*-GLUCITOL MONOOCTADECANOATE ✦ ANHYDROSORBITOL STEARATE ✦ ARLACEL ✦ ARMOTAN ✦ CRILL ✦ DREWSORB ✦ HODAG ✦ IONET ✦ LIPOSORB S-20 ✦ MONTANE ✦ NEWCOL ✦ NIKKOL ✦ SORBITAN C ✦ SORBITAN MONOOCTADECANOATE ✦ SORBITAN STEARATE

SORBITOL

Products and Uses: An additive in cosmetic creams, lotions, toothpaste, and tobacco. Used in baked goods, baking mixes, beverages (low calorie), candy (hard and soft), chewing gum, chocolate, coconut (shredded), cough drops, frankfurter (labeled), frozen dairy desserts, jams (commercial nonstandardized), jellies (commercial nonstandardized), knockwurst, sausage (cooked), and wieners. It is an anticaking agent, sweetener, curing agent, moisture-conditioning, drying agent, emulsifier (stabilizes and maintains mixes), firming agent, flavoring agent, formulation aid, free-flow agent, humectant (moisturizer), lubricant, sweetener (nutritive), pickling agent, release agent, sequestrant (affects the final product's appearance, flavor or texture), stabilizer (helps to keep uniform consistency), surface-finishing agent, texturizing (smoothing) agent, and thickening agent.

Precautions: Mildly toxic by swallowing. Effect on the human body by swallowing is diarrhea. FDA states GRAS (generally recognized as safe) when used within limitations. Diabetics find it useful as a result of slow absorption because blood sugar does not increase rapidly. It may have a laxative effect on some people.

Synonyms: CAS: 50-70-4 ✦ CHOLAXINE ✦ DIAKARMON ✦ GLUCITOL ✦ *d*-GLUCITOL ✦ GULITOL ✦ *l*-GULITOL ✦ KARION ✦ NIVITIN ✦ SIONIT ✦ SIONON ✦ SORBICOLAN ✦ SORBITE ✦ *d*-SORBITOL ✦ SORBO ✦ SORBOL ✦ SORBOSTYL ✦ SORVILANDE

SPEARMINT

Products and Uses: Derived from a plant leaf and used in ice cream, candies, bakery products, spices, gum, condiments, jellies, beverages, perfumes, cosmetics, and toothpaste. It is a flavoring and odorant.

Precautions: Mildly toxic by swallowing. A mutagen (changes inherited characteristics). A skin irritant and an allergen. FDA states GRAS (generally recognized as safe).

Synonyms: CAS: 8008-79-5 ✦ OIL OF SPEARMINT ✦ COMMON SPEARMINT ✦ SCOTCH SPEARMINT ✦ GARDEN MINT ✦ GREEN MINT

SPIKE LAVENDER OIL

Products and Uses: A natural ingredient added to beverages, ice creams, candies, bakery products; also in perfumes, shaving lotions, soaps, and colognes. It is a seasoning or scenting agent.

Precautions: Moderately toxic by swallowing. A skin irritant. FDA states GRAS (generally recognized as safe) when used at moderate levels to accomplish the intended effect.

Synonyms: CAS: 84837-04-7 ✦ LAVENDER OIL, SPIKE ✦ OIL OF SPIKE LAVENDER

SQUALANE

Products and Uses: Found in pharmaceuticals, over-the-counter drug products, cosmetics, and perfumes. It is a synthetic lubricating oil and vehicle base for consumer products. It prevents drying and congealing of products.

Precautions: Could cause allergic reaction in susceptible individuals.

Synonyms: CAS: 111-01-3 ✦ PERHYDROSQUALENE ✦ SPINACANE ✦ DODECAHYDROSQUALANE ✦ HEXAMETHYLTETRACOSANE

SQUALENE

Products and Uses: A natural raw material found in human sebum (skin oil) and in shark liver oil. Used as a lubricating oil in perfume and hair products.

Precautions: Could cause allergic reaction to susceptible individuals.

Synonyms: CAS: 7683-64-9 ✦ SPINACENE

STANNIC CHLORIDE

Products and Uses: An ingredient in metallic hair dyes, soaps, and perfumes. Used for bacteria and fungi control; for perfume stabilizer.

Precautions: Moderately toxic by breathing. A corrosive irritant to skin, eyes, nose and throat.

Synonyms: CAS: 7646-78-8 ✦ TIN CHLORIDE ✦ TIN PERCHLORIDE ✦ TIN TETRACHLORIDE ✦ LIBAVIUS FUMING SPIRIT

STANNIC OXIDE

Products and Uses: A chemical used in fingernail polish, glass polish, putty, perfumes, and cosmetics. It is a mineral used in polishes and in ceramics.

Precautions: Harmless when used for intended purposes.

Synonyms: CAS: 18282-10-5 ✦ CASSITERITE ✦ STANNIC ANHYDRIDE ✦ TIN PEROXIDE ✦ TIN DIOXIDE ✦ STANNIC ACID

STANNOUS FLUORIDE

Products and Uses: A common chemical used in toothpaste to prevent tooth decay.

Precautions: Poison by swallowing. In excessive amounts it is a possible carcinogen (causes cancer) and mutagen (changes inherited characteristics).

Synonyms: CAS: 7783-47-3 ✦ TIN FLOURIDE ✦ FLUORISTAN ✦ STANNOUS FLOURIDE ✦ TIN BIFLUORIDE ✦ TIN DIFLUORIDE

STAPLE FIBER ACETATE

Products and Uses: Used in cigarette filter tips, tobacco smoke filters, and tobacco smoke filter tip rods.

Precautions: A component of cigarettes. Smoking is a recognized cause of cancer, lung disease, emphysema, and heart problems.

Synonyms: COTTON FIBER

STARCH

Products and Uses: A substance used in adhesives (gummed paper, tapes, cartons, bags), food products (gravies, custards, confectionery), filler in baking powders (cornstarch), laundry starch, over-the-counter pharmaceuticals, dentifrices, hair colorings, facial rouge, cosmetics, baby powders, dusting powder, and face powder. It is a gelling agent, sizing agent, fabric stiffener, thickener, anticaking agent, and used in explosives (nitrostarch).

Precautions: Could cause an allergic reaction. Conversely, it is sometimes used to treat skin irritation.

Synonyms: CAS: 9005-84-9 ✦ CARBOHYDRATE POLYMER

STEARIC ACID

Products and Uses: A common ingredient used as a defoaming agent, softener, and flavoring agent. Used in bar soaps, deodorants, cosmetics, facial makeup, hand lotions, ointments, shaving creams, shoe polish, automobile metal polish, skin moisturizers, and suppositories. It is also used in beverages, bakery products, gum, and candies for moisturizing, lubricating, and for producing pearlized effect in hand creams and soaps.

Precautions: A human skin irritant. Can cause allergic reaction in susceptible individuals. FDA states GRAS (generally recognized as safe) when used within specifications.

Synonyms: CAS: 57-11-4 ✦ CENTURY ✦ DAR-CHEM ✦ EMERSOL ✦ GLYCON ✦ GROCO ✦ 1-HEPTADECANECARBOXYLIC ACID ✦ HYDROFOL ACID ✦ HY-PHI ✦ HYSTRENE ✦ INDUSTRENE ✦ KAM ✦ NEO-FAT ✦ OCTADECANOIC ACID ✦ PEARL STEARIC ✦ STEAREX BEADS ✦ STEAROPHANIC ACID ✦ TEGOSTEARIC

STEROIDS, ANABOLIC

Products and Uses: Utilized in drugs and medications to promote growth and repair body tissue.

Precautions: A possible carcinogenic (causes cancer). May be teratogenic (abnormal fetal development) and causes reproductive (infertility, or sterility, or birth defects) effects. Anabolic steroids are synthetic derivatives of testosterone (male hormone). They have beneficial medical benefits when used appropriately for senility, debilitating illness, and for certain convalescents. They are much abused by athletes and can cause serious harmful effects.

Synonyms: VARIOUS

STRYCHNINE

Products and Uses: Used as a rat and mouse poison.

Precautions: Lethal human poison. Effects on the human body by swallowing are muscular twitching, convulsions, spasms, and death. On EPA's Extremely Hazardous Substances list.

Synonyms: CAS: 57-24-9 ✦ CERTOX ✦ KWIK-KIL ✦ RO-DEX ✦ SANASEED ✦ STRYCHNOS ✦ MOUSE-TOX ✦ MOUSE-RID

SUCCINIC ACID

Products and Uses: This chemical is a flavor enhancer and neutralizing agent that eliminates both acidity and alkalinity in order to keep the product neutral. It is used in beverages, condiments, meat products, relishes, sausage (hot), mouth wash, perfumes, dyes, and lacquers.

Precautions: A severe eye irritant. FDA states GRAS (generally recognized as safe) when used within limitations.

Synonyms: CAS: 110-15-6 ✦ AMBER ACID ✦ BERNSTEINSAURE (GERMAN) ✦ BUTANEDIOIC ACID ✦ 1,2-ETHANEDICARBOXYLIC ACID ✦ ETHYLENESUCCINIC ACID

SUCROSE

Products and Uses: A table sugar that is available as granulated, brown, and powdered sugar. Used as a sweetener in desserts, beverages, cakes, ice creams, icings, cereals, and baked goods.

Precautions: Mildly toxic by swallowing. GRAS (generally recognized as safe) when used at moderate levels. Provides no vitamins, minerals, or protein. Americans consume over 60 pounds per person each year.

Synonyms: CAS: 57-50-1 ✦ BEET SUGAR ✦ CANE SUGAR ✦ CONFECTIONER'S SUGAR ✦ α-d-GLUCOPYRANOSYL β-d-FRUCTOFURANOSIDE ✦ (α-d-GLUCOSIDO)-β-d-FRUCTOFURANOSIDE ✦ GRANULATED SUGAR ✦ ROCK CANDY ✦ SACCHAROSE ✦ SACCHARUM ✦ SUGAR

SULFAMIC ACID

Products and Uses: A stabilizing additive in swimming pool chemicals; also used for flameproofing of fabrics and wood, as a weed killer, and in food packaging.

Precautions: Moderately toxic by swallowing. A human skin irritant. A corrosive irritant to skin, eyes, nose and throat. A substance that migrates to food from packaging materials. FDA states GRAS (generally recognized as safe) when used appropriately.

Synonyms: CAS: 5329-14-6 ✦ AMIDOSULFONIC ACID ✦ AMIDOSULFURIC ACID ✦ AMINOSULFONIC ACID ✦ SULFAMIDIC ACID ✦ SULPHAMIC ACID

SULFITES

Products and Uses: Used in various food and beverage products as a preservative and antioxidant (slows down the spoiling of foods, prevents rancidity).

Precautions: Banned from restaurant and supermarket use on salads and vegetables as a result of reactions of individuals. These included unconsciousness, anaphylactic shock, nausea, diarrhea, and asthma attacks, which resulted in deaths in some incidents. It is still used in wines and packaged foods when appropriately labeled.

Synonyms: SULFUR DIOXIDE ✦ SODIUM BISULFITE ✦ SODIUM METABISULFITE ✦ POTASSIUM METABISULFITE

SULFORAPHANE

Products and Uses: A chemical that triggers enzymes known to neutralize carcinogens in cells. Primarily found in cruciferous vegetables, broccoli, brussels sprouts, cabbage, and cauliflower.

Precautions: Very beneficial as an anticancer ingredient.

Synonyms: NONE KNOWN.

SULFURIC ACID

Products and Uses: Battery acid.

Precautions: (EXTREME CAUTION SHOULD BE EXERCISED WHEN CHARGING BATTERIES OR JUMP-STARTING VEHICLES). Human poison by unspecified route. Moderately toxic by swallowing. A severe eye irritant. Extremely irritating, corrosive, and toxic to tissue, resulting in destruction of tissue, causing severe burns. Repeated contact with dilute solutions can cause a dermatitis, and repeated or prolonged breathing can cause inflammation of the upper respiratory tract leading to chronic bronchitis. Exposure may cause a chemical pneumonia.

Synonyms: CAS: 7664-93-9 ✦ VITRIOL BROWN OIL ✦ OIL OF VITRIOL ✦ HYDROOT ✦ DIPPING ACID ✦ NORDHAUSEN ACID ✦ MATTING ACID ✦ SULPHURIC ACID

SULFUROUS ACID

Products and Uses: A fruit, nut, food, and wine preservative. Also in dental bleaching agents. Used as an antiseptic, preservative, and bleach.

Precautions: Toxic by swallowing and breathing.

Synonyms: CAS: 7782-99-2 ✦ 6% SULFUR DIOXIDE SOLUTION

SUNFLOWER OIL

Products and Uses: An ingredient in margarine, shortening, soaps, and dietary supplements. Commonly used as a coating agent, emulsifying agent (stabilizes and maintains mixes), formulation aid, and texturizing agent.

Precautions: Harmless when used for intended purposes.

Synonyms: NONE FOUND.

SUNSCREENS

Products and Uses: Products in the form of lotions, oils, gels, creams, mousse, sticks, and roll-ons used to protect the skin from sun exposure and damage. It is practical to use a high SPF (sun protection factor) numbered product.

Precautions: No sunscreen can completely block out the sun's damaging rays. Therefore, it is necessary to limit exposure and reapply after swimming and exercise. The sun's two ultraviolet rays are UVA and UVB. UVA primarily causes premature aging and wrinkling; UVB primarily causes sunburn and skin cancer, although both can cause aging and cancer.

Synonyms: OXYBENZONE ✦ DIOXYBENZONE ✦ EUSOLAX 6300 ✦ EUSOLAX 8020 ✦ PARSOL1789 ✦ SULISOBENZONE ✦ RED PETROLATUM ✦ TITANIUM DIOXIDE

SYMCLOSENE

Products and Uses: A chemical in household cleansers, deodorizers, and cleaning products. Used as a disinfectant, deodorizer, and chlorinating agent.

Precautions: Moderately toxic by swallowing. Effects on the human body by swallowing are ulceration or bleeding from the stomach. A severe eye and skin irritant.

Synonyms: CAS: 87-90-1 ✦ TRICHLORINATED ISOCYANURIC ACID ✦ ISOCYANURIC CHLORIDE ✦ TRICHLOROCYANURIC ACID

SYNTHETIC PARAFFIN

Products and Uses: Frequently found on grapefruit, lemons, limes, muskmelons, oranges, sweet potatoes, and tangerines as a coating (protective).

Precautions: FDA approves use at moderate levels to accomplish the intended effect.

Synonyms: SUCCINIC DERIVATIVES

TAGETES

Products and Uses: A yellow coloring derived from flower petals of the Aztec marigold. Used as an additive to chicken feed to increase yellow color of the skin and eggs of poultry.

Precautions: Harmless when used for intended purposes although some sensitive individuals could show allergic reactions.

Synonyms: *TAGETES ERECTA L*

TALC

Products and Uses: A product used in baby, bath, foot, and face powders. Commonly found in pharmaceuticals, soaps, cosmetics, eye shadow, rouges, facial makeup, and creams. In paints, putty, lubricants, slate pencils, and crayons. Also used as an anticaking agent, coating agent, lubricant, release agent, surface-finishing agent, and texturizing agent.

Precautions: Could be an allergen and a skin irritant. Prolonged or repeated breathing can produce a form of pulmonary fibrosis (talc pneumoconiosis) that may be due to asbestos in powder. A possible carcinogen. A study of women who used talcum powder on sanitary napkins indicated an increased risk of ovarian cancer. The study indicates that talc entered the reproductive tract in this manner.

Synonyms: CAS: 14807-96-6 ✦ AGALITE ✦ AGI TALC ✦ ALPINE TALC USP ✦ ASBESTINE ✦ DESERTALC ✦ FIBRENE ✦ LO MICRON ✦ METRO TALC ✦ MISTRON FROST ✦ MISTRON STAR ✦ MISTRON SUPER FROST ✦ MISTRON VAPOR ✦ PURTALC USP ✦ SIERRA C-400 ✦ SNOWGOOSE ✦ SOAPSTONE ✦ STEAWHITE ✦ STEATITE ✦ SUPREME DENSE ✦ TALCUM

TALL OIL

Products and Uses: An oil used as a lubricant, paint vehicle, and emulsifier in drying oil, soaps, greases, and paints.

Precautions: A mild allergen. Derived from pine wood. GRAS (generally recognized as safe). An indirect additive from packages. A substance that migrates to food from packaging materials.

Synonyms: CAS: 8002-26-4 ✦ LIQUID ROSIN ✦ TALLOL

TALLOW

Products and Uses: The substance used in soap stock, leather dressing, candles, greases, and in animal feeds. It is useful as a coating agent, emulsifying (stabilizes and maintains mixes) agent, formulation aid, and as a texturizer. Derived from animal (cattle) fat.

Precautions: GRAS (generally recognized as safe) when used for intended purposes.

Synonyms: BLEACHED-DEODORIZED TALLOW

TANGERINE OIL

Products and Uses: Derived from the tangerine fruit and added in desserts, soft drinks, ice cream, and furniture polish; also an odorant in perfume and cosmetics.

Precautions: A skin irritant. GRAS (generally recognized as safe) when used at a level not in excess of the amount necessary to accomplish the desired results.

Synonyms: CAS: 8008-31-9 ✦ TANGERINE OIL, COLDPRESSED (FCC) ✦ TANGERINE OIL, EXPRESSESED (FCC) ✦ CITRUS PEEL OIL

TANNIC ACID

Products and Uses: Derived from tree barks, nutgalls, and plant parts. It is used as a clarifying agent in beer and wine. Used as a lubricant, paint vehicle, and emulsifier. Found in apple juice, baked goods, beer, candy (hard and soft), cough drops, fats (rendered), fillings, frozen dairy desserts, and mixes, gelatins, meat products, nonalcoholic beverages, puddings, and wine. Also used in inks, pharmaceuticals, astringents, and in the treatment of minor burns.

Precautions: Poison by swallowing. FDA states GRAS (generally recognized as safe) when used within limitations.

Synonyms: CAS: 1401-55-4 ✦ D'ACIDE TANNIQUE (FRENCH) ✦ GALLOTANNIC ACID ✦ GALLOTANNIN ✦ GLYCERITE ✦ TANNIN

TARRAGON OIL

Products and Uses: A seasoning oil for liqueurs, salad dressings, soups, and sauces; also an ingredient in perfumes.

Precautions: Moderately toxic by swallowing. A skin irritant. FDA states GRAS (generally recognized as safe) when used within reasonable limits.

Synonyms: CAS: 8016-88-4 ✦ ESTRAGON OIL

TARTARIC ACID

Products and Uses: In baking powder, beverages (grape and lime flavored), jellies (grape flavored), poultry, and wine. It is an acid, firming agent, flavor enhancer, flavoring agent, humectant (keeps product from drying out), sequestrant (affects the product's appearance, flavor or texture).

Precautions: Mildly toxic by swallowing. FDA states GRAS (generally recognized as safe) when used within stated limitations.

Synonyms: CAS: 87-69-4 ✦ 2,3-DIHYDROSUCCINIC ACID ✦ 2,3-DIHYDROXYBUTANEDIOC ACID

TCSA

Products and Uses: Utilized as a bacteriostat (kills germs) to prevent bacterial growth in cleaning products. Found in surgical soaps, laundry soaps, rinses, deodorants, shampoos, and polishes.

Precautions: Poison by swallowing. A skin irritant.

Synonyms: 3,3',4',5-TETRACHLOROSALICYLANILDE ✦ 3,5-DICHLORO-n-(3,4-DICHLOROPHENYL)-2-HYDROBENZAMIDE ✦ IRGASAN BS2000

TEA

Products and Uses: A hot or cold brewed drink used as a refreshment or stimulant beverage.

Precautions: Polyphenol substances in tea are believed to inhibit the action of carcinogens in food and tobacco, and may protect against heart disease. Other elements of tea may prevent tooth decay. Tannins in tea help fight bacteria and viruses.

Synonyms: GREEN TEA ✦ OOLONG TEA ✦ BLACK TEA ✦ HERBAL TEAS ✦ GINSENG TEA

TEA LAURYL SULFATE

Products and Uses: Utilized as a foaming, wetting, and dispersing agent in detergents, cosmetics, shampoos, and pharmaceuticals.

Precautions: A skin and eye irritant.

Synonyms: CAS: 139-96-8 ✦ TRIETHANOLAMINE LAURYL SULFATE ✦ SULFURIC ACID, MONODECYL ESTER ✦ SULFURIC ACID, DODECYL ESTER, TRIETHANOLAMINE SALT

TERPINEOL

Products and Uses: An ingredient in perfumes, soaps, antiseptics, disinfectants, and flavoring agents.

Precautions: Mildly toxic by swallowing. A skin irritant.

Synonyms: CAS: 8006-39-1 ✦ p-MENTH-1-EN-8-OL ✦ MIXTURE OF p-METHENOLS ✦ α-TERPINEOLS

THIOLACTIC ACID

Products and Uses: The active chemical in depilatories (hair removers), permanent wave solutions, and hair-straightening products. Used in hair-curling, removing, and processing products.

Precautions: Poisoning by swallowing. Mildly toxic by breathing. Can cause severe allergic reaction and skin irritation.

Synonyms: CAS: 79-42-5 ✦ 2-MERCAPTOPROPIONIC ACID ✦ 2-MERCAPTOPROPANOIC ACID ✦ 2-THIOLPROPIONIC ACID ✦ α-MERCAPTOPROPANOIC ACID

THIOUREA

Products and Uses: Found in photocopy paper, photography, dyes, drugs, and hair preparations as a fixer; also used as a mold and mildew preventative.

Precautions: A probable carcinogen and a poison. A human mutagen (changes inherited characteristics). Effects on the human body by swallowing are hemorrhage, blood cell changes, and possible depression of bone marrow with anemia. May cause allergic skin eruptions.

Synonyms: CAS: 62-56-6 ✦ THIOCARBAMIDE

THYMOL

Products and Uses: An antibacterial and antifungal drug used in perfumes, deworming medications; a mold and mildew preventative and preservative.

Precautions: Poison by swallowing. Could cause allergic reaction. An FDA over-the-counter drug.

Synonyms: CAS: 89-83-8 ✦ ISOPROPYL-m-CRESOL ✦ THYME CAMPHOR ✦ THYMIC ACID ✦ 3-p-CYMENOL

TITANIUM DIOXIDE

Products and Uses: A white pigment produced from the lightweight silver gray metal ore combined with oxygen. This white pigment is used in paints, slick papers, eye shadows, and picnic forks. It is also the ingredient found in candy, creamed-type canned products, ham salad spread (canned), icings, poultry salads, poultry spreads, and sugar syrup. Micronized titanium dioxide is used in the newest sunscreens. It reflects ultraviolet rays instead of absorbing rays as traditional zinc oxide-based sunscreens do. Commonly used as a color additive, a filler, sunscreen, and pigment.

Precautions: A possible skin irritant. FDA and USDA approve use within limitations.

Synonyms: CAS: 13463-67-7 ✦ A-FIL CREAM ✦ ATLAS WHITE TITANIUM DIOXIDE ✦ AUSTIOX ✦ BAYERITIAN ✦ BAYERTITAN ✦ BAYTITAN ✦ CALCOTONE WHITE T ✦ C.I. PIGMENT WHITE ✦ COSMETIC WHITE C47-5175 ✦ C-WEISS 7 (GERMAN) ✦ FLAMENCO ✦ HOMBITAN ✦ HORSE HEAD A-410 ✦ KRONOS TITANIUM DIOXIDE ✦ LEVANOX WHITE RKB ✦ RAYOX ✦ RUNA RH20 ✦ RUTILE ✦ TIOFINE ✦ TIOXIDE ✦ TITANIC ANHYDRIDE ✦ TITANIC ACID ANHYDRIDE ✦

TITANIUM OXIDE ✦ TITANIUM WHITE ✦ TITANIA ✦ TRIOXIDE(S) ✦ TRONOX ✦ UNITANE O-110 ✦ ZOPAQUE

TOCOPHEROLS

Products and Uses: One of the compounds that compose vitamin E. It protects human cells and fatty tissues from free radicals, reactive oxygen molecules that damage cell membranes and DNA. They also protect against a decline in immune response as the body ages.

Precautions: Research indicates 200 mg a day represent the optimal level of vitamin E for immune response.

Synonyms: NONE KNOWN.

TOLUENE

Products and Uses: Derived from coal tar. Used as a thinner and solvent. Various chemical uses include in paint, solvents, lacquers, art supplies, nail polish, cosmetics, dry-cleaning products, spot removers, waxes, gasoline, detergents, dyes, pharmaceuticals, perfumes, and adhesive solvent for plastic toys and novelties.

Precautions: Mildly toxic by breathing. Central nervous system (CNS) effects on the human body by breathing are hallucinations, distorted perceptions, motor activity changes, psychophysiological test changes, and bone marrow changes. An eye irritant.

Synonyms: CAS: 108-88-3 ✦ ANTISAL 1a ✦ BENZENE, METHYL- ✦ METHACIDE ✦ METHANE, PHENYL- ✦ METHYLBENZENE ✦ METHYLBENZOL ✦ PHENYLMETHANE ✦ TOLUEEN (DUTCH) ✦ TOLUEN (CZECH) ✦ TOLUOL (DOT) ✦ TOLUOLO (ITALIAN) ✦ TOLU-SOL

TONKA

Products and Uses: Derived from seeds of fruit of tree *Dipteyx Odorata*. It is used in flavorings, extracts, powders, deodorants, detergents, shampoos, soaps, and antiseptics. Also used as deodorizing and odor-enhancing agent; in pharmaceutical preparations.

Precautions: Moderately toxic by swallowing. A skin irritant.

Synonyms: CAS: 91-64-5 ✦ COUMARIN ✦ CUMARIN ✦ BENZOPYRONE ✦ TONKA BEAN CAMPHOR

TRAGACANTH GUM

Products and Uses: A common ingredient found in baked goods, beverages (citrus), condiments, fats, fruit fillings, gravies, meat products, oil, relishes, salad dressings, and sauces. In adhesives, leather dressing, toothpaste, soap chip coating, soap powders, hairwave preparations, printing inks, and medication tablet binders (holds components together). As an emulsifier (stabilizes and maintains mixes), preservative, and thickening agent.

Precautions: Mildly toxic by swallowing. A mild allergen. FDA states GRAS (generally recognized as safe) when used within stated limits.

Synonyms: CAS: 9000-65-1 ✦ GUM TRAGACANTH ✦ TRAGACANTH

TRICHLOROETHYLENE

Products and Uses: The primary component in dry-cleaning and spot-removing products. Also used in small amounts in cosmetics, perfumes, after-shaves, and cosmetics. Used as a solvent, dye, adhesive, paint, and cleaner.

Precautions: A suspected carcinogen (causes cancer). Toxic by breathing and swallowing. Effects on the body by swallowing are sleepiness, hallucinations, vision problems, confusion, gastrointestinal (stomach) problems, and jaundice (yellowing of skin and eyes). A human mutagen (changes inherited characteristics). A severe skin and eye irritant. Breathing causes headache, drowsiness, and possibly unconsciousness. Can cause cardiac failure.

Synonyms: CAS: 79-01-6 ✦ ACETYLENE TRICHLORIDE ✦ BENZINOL ✦ ETHINYL TRICHLORIDE ✦ EHYLENE TRICHLORIDE ✦ TRICHLORAN ✦ TRICHLORETHENE ✦ THRETHYLENE

TRICHLOROISOCYANIC ACID

Products and Uses: Frequently found in household dry bleaches, dishwashing compounds, scouring powders, detergent sanitizers, laundry bleaches, swimming pool, hot tub, and jacuzzi disinfectants. Used as a bactericide, algicide, bleach, and deodorant.

Precautions: Moderately toxic by swallowing. Effects on the body from swallowing include ulceration or bleeding from the stomach. A severe skin and eye irritant.

Synonyms: CAS: 87-90-1 ✦ ISOCYANIC CHLORIDE ✦ SYMCLOSEN ✦
TRICHLORINATED ISOCYANURIC ACID ✦ TRICHLOROISOCYANURIC ACID

TRICHLOROFLUORMETHANE

Products and Uses: The chemical used for fire extinguishers, refrigerators, and solvents. It is an aerosol propellant and refrigerant.

Precautions: High concentrations cause numbness and anesthesia. Effects on the body by breathing include eye irritation, lung and liver changes.

Synonyms: CAS: 75-69-4 ✦ FLUOROTRICHLOROMETHANE ✦
FLUOROCARBON-11

TRICLOSAN

Products and Uses: The chemical added to deodorants, hand soaps, vaginal deodorants, cosmetics, pharmaceuticals, acne medications, surgeons' antimicrobials, and cleansers. A disinfectant and antibacterial common in consumer toiletries. It is also being added to plastic cutting boards and children's toys to kill bacteria. The compound is added during the extrusion of plastics and fibers for making cloths. During the meltdown process, the triclosan chemically bonds with the materials. It does not react with plastic but is left sitting in the gaps between the polymer chains.

Precautions: Moderately toxic by swallowing. Mildly toxic by skin contact. A human mutagen (changes inherited characteristics). A skin irritant and possible allergen.

Synonyms: CAS: 3380-34-5 ✦ IRGASAN ✦
5-CHLORO-2-(2,4-DICHLOROPHENOXYPHENOL) ✦
2'-HYDROXY-2.4.4'-TRICHLORO-PHENETHYLETHER

TRIETHANOLAMINE

Products and Uses: An ingredient in dry-cleaning soaps, shaving soaps, shampoos, cosmetics, household detergents, fruit coatings, and vegetable coatings. It is a softening agent, emulsifier (stabilizes and maintains mixes), and humectant (keeps product from drying out).

Precautions: Mildly toxic by swallowing. A skin and eye irritant.

Synonyms: CAS: 102-71-6 ✦ TRIHYDROXYTRIETHYLAMINE ✦ STEROLAMIDE ✦ TRIETHYLOLAMINE ✦ TROLAMINE ✦ DALTOGEN ✦ TEA

TRIPALMITIN

Products and Uses: A common additive used in shampoos, soaps, shaving creams, and leather polish. It is a texturizer (smooths ingredients) in toiletries and consumer products.

Precautions: Harmless when used for intended purposes.

Synonyms: CAS: 555-44-2 ✦ PALMITIN ✦ GLYCERYL TRIPALMITATE ✦ HEXADECANOIC ACID

TRIPOLI

Products and Uses: The term for an ingredient in scouring soaps, scouring powders, polishing powders, paints, and wood filler. Used as an abrasive, absorbent, polish, filler, base, and filter.

Precautions: The prolonged breathing of dusts may cause disabling lung problem known as silicosis.

Synonyms: CAS: 1317-95-9 ✦ ROTTENSTONE ✦ AMORPHOUS SILICA ✦ SILICA

TRISODIUM PHOSPHATE

Products and Uses: A chemical in water softeners, detergents, dishwashing compounds, and scouring powders. Used as a conditioner, cleaner, and bactericide.

Precautions: Irritant to skin and eyes.

Synonyms: CAS: 7601-54-9 ✦ SODIUM PHOSPHATE, TRIBASIC ✦ PHOSPHORIC ACID, TRISODIUM SALT ✦ SODIUM PHOSPHATE ✦ TRISODIUM ORTHOPHOSPHATE ✦ TRISODIUM PHOSPHATE ✦ TROMETE ✦ TSP

TRISULFURATED PHOSPHORUS

Products and Uses: A chemical in match tips. It is a flammable solid.

Precautions: Poison by swallowing.

Synonyms: CAS: 1314-85-8 ✦ TETRAPHOSPHORUS TRISULFIDE ✦ PHOSPHORUS SESQUISULFIDE

TROMETHAMINE

Products and Uses: A substance added to leather polishes, cleaning polishes, skin lotions, pharmaceuticals, cosmetics, and creams. It is an emulsifier (stabilizes and maintains mixes) in polishes and cleaning compounds.

Precautions: Moderately toxic by swallowing.

Synonyms: CAS: 77-86-1 ✦ THAM ✦ PEHANORM ✦ TALATROL ✦ TRIMETHYLOL AMINOMETHANE ✦ TRISAMINE ✦ TRIS BUFFER ✦ TROMETAMOL ✦ TROMETHANE ✦ TALALTROL

TSPP

Products and Uses: A chemical in shampoos, cleaning compounds, water softeners, rust stain removers, spot removers, and ink removers. Frequently used as a sequestrant (binds ingredients that affect the final product's appearance, flavor or texture), clarifying agent, and buffering (regulates acidity and alkalinity) agent.

Precautions: Poison by swallowing.

Synonyms: CAS: 7722-88-5 ✦ PYROPHOSPHATE ✦ TETRASODIUM PYROPHOSPHATE ✦ SODIUM PYROPHOSPHATE ✦ TETRASODIUM DIPHOSPHATE ✦ PHOSPHOTEX

TUNG NUT OIL

Products and Uses: Added to exterior paints and varnishes. Derived from seeds of plant grown in the Orient. Used for waterproofing, finishes, and packaging materials.

Precautions: A skin irritant. Toxic by swallowing. Causes nausea, vomiting, cramps, diarrhea, rectal spasms, thirst, dizziness, sleepiness, and confusion. Large doses can cause fever, disturbance of heart rhythms, and breathing effects.

Synonyms: CHINAWOOD OIL

TURPENTINE

Products and Uses: An oily liquid derived from the pine tree. It is in varnishes, insecticides, paint thinners, wax-based polishes, shoe polishes, furniture polishes, liniments (medicinal), antiseptics, and perfumery. A common solvent.

Precautions: Moderately toxic by swallowing. Effects on the human body by swallowing and breathing are nasal and eye irritation, hallucinations, confusion, headache, and lung and kidney damage. Irritating to skin, nose, and throat. A very dangerous fire hazard when exposed to heat or flames.

Synonyms: CAS: 8006-64-2 ✦ OIL OF TURPENTINE ✦ SPIRIT OF TURPENTINE ✦ TURPENTINE OIL

ULTRAMARINE BLUE

Products and Uses: A laundry bluing used to offset yellow tones. Also added in soaps and cosmetics (eye shadows and mascaras). Used as a coloring agent. It occurs naturally as the blue pigmentation in the mineral lapis lazuli.

Precautions: Not harmful for external use.

Synonyms: CAS: 57455-37-5 ✦ C.I. PIGMENT BLUE

UMBELLIFERONE

Products and Uses: Found in cosmetics and sunscreen cream, lotion, gel, and spray preparations. Used as an ultraviolet (UV) screening agent.

Precautions: Harmless when used for intended purposes.

Synonyms: CAS: 93-35-6 ✦ 7-HYDROXYCOUMARIN ✦ HYDRANGIN ✦ SKIMMETIN

γ-UNDECALACTONE

Products and Uses: An ingredient used to affect the taste and smell of beverages, candy, gelatins, ice cream, puddings, and perfumes. Provides a peachlike odor.

Precautions: Harmless when used for intended purposes.

Synonyms: ALDEHYDE C-14 PURE ✦ FEMA NO. 3091 ✦ PEACH ALDEHYDE

UNDECANAL

Products and Uses: A chemical in desserts, ice creams, bakery products, candies, and chewing gum. It is used for fruit and floral flavorings and scents.

Precautions: A skin irritant.

Synonyms: CAS: 112-44-7 ✦ ALDEHYDE-14 ✦ 1-DECYL ALDEHYDE ✦ HENDECANAL ✦ HENDECANALDEHYDE ✦ UNDECANAL ✦ n-UNDECANAL ✦ UNDECANALDEHYDE ✦ UNDECYL ALDEHYDE ✦ n-UNDECYL ALDEHYDE ✦ UNDECYLIC ALDEHYDE

UNDECYL ALCOHOL

Products and Uses: Added to beverages, ice creams, candies, and bakery products. Used as a synthetic citrus and floral flavoring.

Precautions: Moderately toxic by swallowing. A skin irritant.

Synonyms: CAS: 112-42-5 ✦ ALCOHOL C-11 ✦ HENDECANOIC ALCOHOL ✦ 1-HENDECANOL ✦ HENDECYL ALCOHOL ✦ n-HENDECYLENIC ALCOHOL ✦ n-UNDECANOL

UREA

Products and Uses: Utilized in alcoholic beverages, gelatin products, wine, yeast-raised bakery products for browning; plant fertilizer and animal feeds. It is present in personal care products, deodorants, toothpastes, mouthwashes, and hair products. Used as a fermentation aid, formulation aid, antiseptic, adhesive, and flameproofer.

Precautions: A human mutagen (changes inherited characteristics). A skin irritant. GRAS (generally recognized as safe) when used within limits.

Synonyms: CAS: 57-13-6 ✦ CARBAMIDE ✦ CARBAMIDE RESIN ✦ CARBAMIMIDIC ACID ✦ CARBONYL DIAMIDE ✦ CARBONYLDIAMINE ✦ ISOUREA ✦ PRESPERSION, 75 UREA ✦ PSEUDOUREA ✦ SUPERCEL 3000 ✦ UREAPHIL ✦ UREOPHIL ✦ UREVERT ✦ VARIOFORM II

UREA PEROXIDE

Products and Uses: Frequently used in cosmetics and pharmaceuticals as a bleaching disinfectant.

Precautions: An irritant.

Synonyms: CAS: 124-43-6 ✦ HYDROGEN PEROXIDE CARBAMIDE ✦ EXTEROL ✦ ORTIZON ✦ HYPEROL ✦ PERHYDRIT ✦ PERHYDROL UREA

URETHANE

Products and Uses: Found in over 1000 beverages sold in the U.S. In sherries, fruit brandies, whiskeys, table wines, dessert wines, liqueurs. A natural product of the fermentation process by which yeast turns fruit juice into beverages. The allowable limit is 125 ppb; before this limit was set, some fruit brandies had 1000 to 12,000 ppb urethane. Used as an intermediate in the manufacture of pharmaceuticals, pesticides, and fungicides.

Precautions: A definite carcinogen (causes cancer) that is toxic by swallowing. It is a human mutagen (changes inherited characteristics). Can cause CNS (central nervous system) depression, nausea, and vomiting.

Synonyms: CAS: 51-79-6 ✦ ETHYL CARBAMATE ✦ ETHYL URETHANE ✦ LEUCETHANE ✦ LEUCOTHANE ✦ PRACARBAMINE ✦ CARBAMIC ACID, ETHYL ESTER

URUSHIOL

Products and Uses: The oily, toxic components of poison ivy and poison oak. Sometimes used in hyposensitization therapy as an antiallergic.

Precautions: Causes severe allergic dermatitis.

Synonyms: CATECHOL DERIVATIVE MIXTURE ✦ *RHUS TOXICODENDRON*

VALERIAN OIL

Products and Uses: An additive used as a tobacco perfume, industrial odorant, and flavoring.

Precautions: Harmless when used for intended purposes.

Synonyms: PINENE ✦ CAMPHENE ✦ BORNEOL ✦ *VALERIANA OFFICINALIS*

VALERIC ACID

Products and Uses: Usually used in perfumes, pharmaceuticals, beverages, ice creams, candies, and bakery products. It is an ingredient that affects the taste or smell of the product.

Precautions: Moderately toxic by swallowing and breathing. A corrosive irritant to skin, eyes, nose, and throat.

Synonyms: CAS: 109-52-4 ✦ BUTANECARBOXYLIC ACID ✦ 1-BUTANECARBOXYLIC ACID ✦ PENTANOIC ACID ✦ n-PENTANOIC ACID ✦ PROPYLACETIC ACID ✦ VALERIANIC ACID ✦ n-VALERIC ACID

VANILLIN

Products and Uses: A common ingredient in perfumes, candies, beverages, various foods, liqueurs, and pharmaceuticals. It is useful as an ingredient that affects the aroma, taste, or scent of final product.

Precautions: Moderately toxic by swallowing. FDA approves use in moderate levels.

Synonyms: CAS: 121-33-5 ✦ 4-HYDROXY-m-ANISALDEHYDE ✦ 4-HYDROXY-3-METHOXYBENZALDEHYDE ✦ LIOXIN ✦ 3-METHOXY-4-HYDROXYBENZALDEHYDE ✦

METHYLPROTOCATECHUALDEHYDE ✦ VANILLA ✦ VANILLALDEHYDE ✦
VANILLIC ALDEHYDE ✦ p-VANILLIN ✦ ZIMCO

VEGETABLE OIL

Products and Uses: Derived from plants, it is utilized in paints, shortenings, salad dressings, margarine, soaps, cosmetics, lipstick, hair products, and shaving products. Used as a softener, carrier, filler, thickener, or a cleanser.

Precautions: Harmless when used for intended purposes. Some individuals may have allergic reactions as the oil is derived from plant products such as peanut, olive, coconut, linseed, sesame, or cottonseed oil.

Synonyms: SEED OIL ✦ NUT OIL ✦ PLANT OIL ✦ MIXED GLYCERIDES

VERMICULITE

Products and Uses: A clay mineral mined from the earth and used for insulation, sound conditioning, fertilizer additive, plaster, soil conditioner, seed bed for plants, and absorbant for oil spills in the ocean. Useful as an extender filler, nonconductor, aerator, sponge, gypsum board, and leavening material.

Precautions: Considered harmless when used for intended purposes.

Synonyms: HYDRATED MAGNESIUM-IRON-ALUMINUM SILICATE

VITAMIN A

Products and Uses: It occurs abundantly in fish and fish liver oils. It is added to milk, cheese, margarine, ice cream, and baby formulations. Also added to skin ointments and creams. Used as a dietary supplement. Possesses beneficial healing properties for skin.

Precautions: Lack of this vitamin results in poor growth, mental abnormalities, and abnormalities of the craniofacial area and urogenital system. FDA states GRAS (generally recognized as safe) when used at moderate levels to accomplish the intended effect. The recommended adult dose is 5000 units.

Synonyms: CAS: 68-26-8 ✦ AFAXIN ✦ AGIOLAN ✦ ALPHASTEROL ✦ ANATOLA ✦ ANTI-INFECTIVE VITAMIN ✦ ANTIXEROPHTHALMIC VITAMIN ✦ AORAL ✦ APEXOL ✦ AVIBON ✦ AVITOL ✦ BIOSTEROL ✦ CHOCOLA A ✦ 3,7-DIMETHYL-9-(2,6,6-TRIMETHYL-1-CYCLOHEXEN-1-YL)-2,4,6,8-NONATETRAEN-1-OL ✦ DISATABS TABS ✦ DOFSOL ✦ EPITELIOL ✦ HI-A-VITA ✦ LARD FACTOR ✦ MYVPACK ✦ OLEOVITAMIN A ✦

OPHTHALAMIN ✦ PREPALIN ✦ RETINOL ✦ all-trans RETINOL ✦ RETROVITAMIN A ✦ TESTAVOL ✦ VAFLOL ✦ VI-ALPHA ✦ VITAMIN A1 ✦ VITAMIN A1 ALCOHOL ✦ all-trans-VITAMIN A ALCOHOL ✦ VITAVEL-A ✦ VITPEX ✦ VOGAN

VITAMIN E

Products and Uses: Wheat germ oil is a rich source of this fat-soluble compound. It is found in bacon, fats (rendered animal), pork fat (rendered), and poultry. Useful as an antioxidant (added to oil-containing food to prevent it from getting rancid), dietary supplement, nutrient, and preservative.

Precautions: FDA states GRAS (generally recognized as safe) when used within stated limits. Deficiency can cause sterility and muscular dystrophy in animals. Recent studies indicate that it is important in the prevention of heart disease and possibly other diseases.

Synonyms: CAS: 59-02-9 ✦ ALMEFROL ✦ ANTISTERILITY VITAMIN ✦ COVI-OX ✦ DENAMONE ✦ EMIPHEROL ✦ ENDO E ✦ EPHYNAL ✦ EPROLIN ✦ EPSILAN ✦ ESORB ✦ ETAMICAN ✦ ETAVIT ✦ EVION ✦ EVITAMINUM ✦ ILITIA ✦ PHYTOGERMINE ✦ PROFECUNDIN ✦ SPAVIT ✦ SYNTOPHEROL ✦ *d*-α-TOCOPHEROL (FCC) ✦ *dl*-α-TOCOPHEROL (FCC) ✦ (R,R,R)-α-TOCOPHEROL ✦ α-TOCOPHEROL ✦ (2R,4′R,8′R)-α-TOCOPHEROL ✦ TOKOPHARM ✦ 5,7,8-TRIMETHYLTOCOL ✦ VASCUALS ✦ VERROL ✦ VITAPLEX E ✦ VITAYONON ✦ VITEOLIN

VOLATILE ORGANIC COMPOUNDS

Products and Uses: Carbon compounds that react with the atmosphere are in this category. Manufacturing operations are the most common source. They are also in hair sprays, windshield glass cleaner, air fresheners, liquid cleaning products, auto engine degreasers, wood furniture polish, wood floor polish, laundry products, nail polish remover, oven cleaners, hair mousse spray, bath and ceramic cleaners, spray insect repellents, hair-setting gels, and shaving creams.

Precautions: Individuals may notice irritation of the eyes, nose and throat, headaches, mental fatigue, and breathing difficulties after exposure. The VOCs are a major component in the production of smog or ozone. This is the cause of great financial loss because of crop and foliage damage.

Synonyms: VOC ✦ VOLATILE ORGANIC CHEMICALS ✦ FORMALDEHYDE ✦ BENZENE ✦ TCE ✦ ORGANIC VOLATILE CHEMICALS ✦ OVC

WARFARIN

Products and Uses: A rodenticide and medical anticoagulant.

Precautions: It is a poison by swallowing and breathing. Moderately toxic by skin contact. Effects on the human body by swallowing include hemorrhage, ulceration, bleeding from small intestine. Human reproductive effects by swallowing include fetus death and abnormalities at birth. Human teratogenic (abnormal fetal development) effects include abnormal head, face, musculoskeletal, and breathing problems.

Synonyms: CAS: 81-81-2 ✦ ARAB RAT DETH ✦ BRUMIN ✦ COMPOUND 42 ✦ d-CON ✦ CO-RAX ✦ COUMADIN ✦ COUMAFENE ✦ LIQUA-TOX ✦ PROTHROMADIN ✦ RAT-A-WAY ✦ RAT-B-GON ✦ RAT-GARD ✦ SOLFARIN

WASHING SODA

Products and Uses: A cleanser and sanitizer in soaps, detergents, and bleaching preparations.

Precautions: Moderately toxic by swallowing or breathing. A skin and eye irritant.

Synonyms: CAS: 497-19-8 ✦ CARBONIC ACID ✦ DISODIUM SALT ✦ CRYSTOL CARBONATE ✦ DISODIUM CARBONATE ✦ SODA ASH ✦ TRONA

WAX

Products and Uses: Derived from many sources including animal, plant, and synthetic. Varied uses include polishes, carbon paper, floor wax, candles, crayons, sealants, cosmetics, and for protecting food products.

Precautions: Harmless when used for intended purposes.

Synonyms: WAX, CHLORONAPHTHALENE ✦ WAX, MICROCRYSTALLINE ✦ WAX, POLYMETHYLENE

WHEAT GLUTEN

Products and Uses: Derived from the wheat grass. Used in bakery bread items, makeup powders and creams. Commonly added as dough conditioner, formulation aid, nutrient supplement, processing aid, stabilizer, surface-finishing agent, texturizing agent, and thickening agent.

Precautions: FDA states GRAS (generally recognized as safe).

Synonyms: CAS: 8002-80-0

WHEY, DRY

Products and Uses: The product of milk coagulation that is added to products such as beef with barbecue sauce, bratwurst, chili con carne, loaves (nonspecific), pork with barbecue sauce, poultry, sausage, sausage (imitation), soups, and stews. It is used as a binder or extender.

Precautions: FDA states GRAS (generally recognized as safe) when used within stated limits.

Synonyms: DRY WHEY ✦ DRIED WHEY

WHISKEY

Products and Uses: Liquor produced from grains. Considered a beverage or medicine; used as a stimulant, antiseptic, or vasodilator.

Precautions: A noncumulative poison usually harmless in moderate amounts, but may be toxic when habitually taken in large amounts.

Synonyms: CORN WHISKEY ✦ RYE WHISKEY ✦ BARLEY WHISKEY ✦ ALCOHOL ✦ ETHANOL

WHITEWASH

Products and Uses: Wall coatings and paint-type products produced from lime and water mixtures. It is an antiseptic coating for barns and chicken houses

Precautions: Package directions should be carefully observed, using necessary eye and skin protection.

Synonyms: KALSOMINE ✦ HYDRATED LIME SUSPENSION ✦ CALCIUM
CARBONATE SUSPENSION.

WILD CHERRY

Products and Uses: Derived from the bark and stems of the plant. It is used in
food, cosmetics, pharmaceuticals, cough drops, syrups, and expectorants.

Precautions: Harmless when used for intended purposes.

Synonyms: *PRUNUS SEROTINA*

WINE

Products and Uses: A product derived from the fermented juice of grapes or
other fruits or plants.

Precautions: Some studies have indicated that small quantities (four ounces per
day) have been found to be beneficial in lowering cholesterol. Large quantities
over extended periods can lead to alcoholism, liver damage, numbness of the
extremities, brain damage, gastritis, and heart muscle damage. Pregnant
women should avoid alcohol entirely because of the disastrous effects on the
unborn.

Synonyms: ALCOHOLIC BEVERAGE

WOODRUFF

Products and Uses: An herb for flavoring May wine and for aromatic potpourri.

Precautions: Harmless when used for intended purposes.

Synonyms: *ASPERULA ODORATA* LEAVES

XANTHAN GUM

Products and Uses: Frequently added in baked goods, batter or breading mixes, beverages, chili (canned), chili with beans (canned), desserts, fish patés, gravies, jams, jellies, meat patés, milk products, dairy products, pizza topping mixes, poultry, salad dressings, salads (meat), sauces, sauces (meat), and stews (canned or frozen). It is also in cosmetic products for emulsifying (stabilizes and maintains mixes) and stabilizing (keeps a uniform consistency) of product. As a binder, bodying agent, emulsifier, extender, foam stabilizer, stabilizer, suspending agent, and thickening agent.

Precautions: FDA approves use at moderate levels to accomplish the desired results. USDA states use limitations.

Synonyms: CAS: 11138-66-2 ✦ CORN SUGAR GUM ✦ *XANTHOMONAS CAMPESTRIS* ✦ POLYSACCHARIDE GUM

XYLENOL

Products and Uses: An ingredient in disinfectants, lubricants, gasoline, solvents, pharmaceuticals, insecticides, and fungicides.

Precautions: Toxic by swallowing and skin absorption.

Synonyms: CAS: 1300-71-6 ✦ DIMETHYLPHENOL ✦ HYDROXYDIMETHYLBENZENE ✦ DIMETHYLHYDROXYBENZENE

XYLITOL

Products and Uses: Derived from the birch tree or from other wood pulp. It is a nutritive sweetener used in gum, candies, and breath mints. It was recently reported by researchers in Finland that regular doses of the natural sweetener reduced ear infections among children by up to 50%. It also has been recommended for the prevention of tooth decay and canker sores.

Precautions: FDA approves use at moderate levels to accomplish the intended effect. Mildly toxic by swallowing.

Synonyms: CAS: 87-99-0 ✦ KLINIT ✦ XYLITE (SUGAR)

YEAST

Products and Uses: A single-celled fungus, which produces enzymes that change sugar to alcohol and carbon dioxide. Used in the fermentation of sugars, molasses, and cereals for alcohol. It is considered a food supplement and source of vitamins. Used in baked, cooked, brewed food and beverage products.

Precautions: Harmless when used for intended purposes.

Synonyms: BARM ✦ *SACCHAROMYCETACEAE*

YLANG YLANG OIL

Products and Uses: Derived from a plant that grows in the Philippines and added to fragrances, perfumes, and colognes. It is also used in fruit and cola-flavored soft drinks.

Precautions: Harmless when used for intended purposes.

Synonyms: CANANGA OIL ✦ *CANANGA ODORATA*

ZANZIBAR GUM

Products and Uses: Fossil resins found on the island of Zanzibar and the African mainland. Used in varnishes and lacquers and as an ingredient in wood-finishing products.

Precautions: Harmless when used for intended purposes. Label directions on products should be carefully followed.

Synonyms: AFRICAN GUM ✦ ZANZIBAR ISLAND GUM

ZEIN

Products and Uses: An additive in confections, grain, nuts, adhesives, and panned goods. It is used for glaze, paper coating, grease-resistant coating, label varnishes, printing inks, shellac substitute, food coatings, microencapsulation, and as a surface-finishing agent.

Precautions: A corn processing byproduct considered nontoxic and harmless when used for intended purposes. FDA states GRAS (generally recognized as safe).

Synonyms: CAS: 9010-66-6

ZEOLITE

Products and Uses: A chemical added to water softeners, detergent builders, adsorbents and desiccants (drying agents).

Precautions: Container directions must be followed carefully.

Synonyms: ANALCITE ✦ CHABAZITE ✦ HEULANDITE ✦ NATROLITE ✦ STILBITE ✦ THOMOSONITE

ZINC ACETATE

Products and Uses: A versatile chemical used in astringents, antiseptics, emetics, and styptics. Also utilized as a wood preservative, dietary supplements, textile dyes, and feed additives.

Precautions: Harmless when used for intended purposes.

Synonyms: CAS: 557-34-6

ZINC ARSENATE

Products and Uses: An insecticide and wood preservative.

Precautions: A confirmed carcinogen (causes cancer). Toxic by swallowing and breathing.

Synonyms: ZINC ORTHOARSENATE ✦ KOETTIGITE MINERAL

ZINC BACITRACIN

Products and Uses: Useful as an antibacterial, antiseptic ointment, suppositories, throat lozenges, and other pharmaceuticals. Used in assorted medical products as an antimicrobial.

Precautions: Harmless when used for intended purposes.

Synonyms: BACITRACIN ZINC COMPLEX ✦ BACIFERM

ZINC CARBONATE

Products and Uses: A chemical in cosmetics, lotions, pharmaceuticals, ointments, and dusting powders; used as a coloring agent in toiletries and consumer products.

Precautions: Harmless when used for intended purposes.

Synonyms: CAS: 3486-35-9 ✦ SMITHSONITE ✦ ZINCSPAR

ZINC CHLORIDE

Products and Uses: An additive in adhesives, dental cements, deodorants, disinfectants, and taxidermy embalming fluid. On lumber it is used for a preservative and fireproofing. Also used in medicines as an astringent and antiseptic.

Precautions: Poison by swallowing. Lung tissue changes can result from breathing. A corrosive irritant to the skin, eyes, nose, and throat.

Synonyms: CAS: 7646-85-7 ✦ BUTTER OF ZINC ✦ TINNING GLUX ✦ ZINC DICHLORIDE ✦ ZINC MURIATE

ZINC CITRATE

Products and Uses: An oral hygiene product used in toothpaste, dentifrice, breath fresheners, and mouthwashes.

Precautions: Harmless when used for intended purposes.

Synonyms: ZINC CARBONATE CITRIC ACID

ZINC DITHIOAMINE COMPLEX

Products and Uses: A fungicide, rat and mouse poison, deer and rabbit repellent. Used as a pesticide and wildlife deterrent.

Precautions: Toxic by swallowing.

Synonyms: ZINC DIMETHYLDITHIOCARBAMATECYCLOHEXYLAMINE

ZINC FLUOROSILICATE

Products and Uses: An additive in concrete hardener, laundry sour, preservative, and mothproofing agents.

Precautions: Poison by swallowing.

Synonyms: CAS: 16871-71-9 ✦ ZINC SILICOFLUORIDE ✦ ZINC FLUOSILICATE ✦ ZINC HEXAFLUOROSILICATE

ZINC FORMATE

Products and Uses: Found in waterproofing agents, textiles, and antiseptics.

Precautions: Toxic by swallowing.

Synonyms: CAS: 557-41-5

ZINC GLUCONATE TRIHYDRATE LOZENGES

Products and Uses: Lozenges formulated to relieve common cold symptoms. A homeopathic, nonsedating, natural medication.

Precautions: Considered harmless when manufacturer's directions are followed.

Synonyms: ZINCUM GLUCONIUM ✦ COLD-EZE

ZINC OXIDE

Products and Uses: A white pigment in paints, cosmetics, dental cements, white glue, matches, white printing inks, artists paints, and so on. Useful as a coloring agent, antiseptic, astringent, dietary supplement, UV (ultraviolet) absorber, and nutrient.

Precautions: Moderately toxic to humans by swallowing. A skin and eye irritant.

Synonyms: CAS: 1314-13-2 ✦ AMALOX ✦ AZODOX-55 ✦ CALAMINE (spray) ✦ CHINESE WHITE ✦ EMANAY ZINC OXIDE ✦ EMAR ✦ FELLING ZINC OXIDE ✦ FLOWERS OF ZINC ✦ HUBBUCK'S WHITE ✦ K-ZINC ✦ OZIDE ✦ OZLO ✦ PASCO ✦ PERMANENT WHITE ✦ PHILOSOPHER'S WOOL ✦ SNOW WHITE ✦ WHITE SEAL ✦ ZINCITE ✦ ZINCOID ✦ ZINC WHITE

ZINC PROPIONATE

Products and Uses: Utilized for adhesive tape and medical bandaging. Useful as a topical fungicide (kills fungi, molds, and bacteria).

Precautions: Harmless when used for intended purposes.

Synonyms: CAS: 557-28-8

ZINC STEARATE

Products and Uses: A multipurpose chemical in tablets, cosmetics, pharmaceuticals, powders, and ointments. Utilized as a waterproofing agent, in antiseptics, astringents, lacquers, and protective coatings.

Precautions: Breathing can cause lung damage.

Synonyms: CAS: 557-05-1

ZIRCONYL CHLORIDE

Products and Uses: A chemical added to body deodorants, antiperspirants, cosmetic additives. Also used as a water repellents and in topical skin products.

Precautions: Moderately toxic by swallowing. Can cause skin irritation.

Synonyms: CAS: 7699-43-6 ✦ ZIRCONIUM OXYCHLORIDE ✦ BASIC ZIRCONIUM CHLORIDE ✦ ZIRCONIUM CHLORIDE

APPENDIX I

United States Poison Control Centers

ALABAMA

Alabama Poison Center, 408-D Paul Bryant Drive, Tuscaloosa, AL 35401
Emergency Phone: 800-462-0800

Regional Poison Control Center, The Children's Hospital of Alabama, 1600 7th Avenue South, Birmingham, AL 35233-1711
Emergency Phone: 205-939-9201, -9202, 205-933-4050, 800-292-6678 (AL only)

ALASKA

Anchorage Poison Control Center, Providence Hospital Pharmacy, P.O. Box 196604, Anchorage, AK 95516-6604
Emergency Phone: 800-478-3193, 907-261-3193

ARIZONA

Arizona Poison and Drug Information Center, Arizona Health Sciences Center, Rm. #1156, 1501 N. Campbell Avenue, Tucson, AZ 85724
Emergency Phone: 520-626-6016, 800-362-0101 (AZ only)

Samaritan Regional Poison Center, 1111 E. McDowell Road, Ancillary 1, Phoenix, AZ 85006
Emergency Phone: 602-253-3334, 800-362-0101 (AZ only)

ARKANSAS

Arkansas Poison and Drug Information Center, University of Arkansas for Medical Sciences, 4301 West Markham-Slot 522, Little Rock, AR 72205
Emergency Phone: 800-376-4766

CALIFORNIA

California Poison Control System—Fresno Valley, Children's Hospital, 3151 N. Millbrook, Fresno, CA 93703
Emergency Phone: 209-445-1222, 800-346-5922

California Poison Control System—San Diego, UCSD Medical Center, 200 West Arbor Drive, San Diego, CA 92103-8925
Emergency Phone: 619-543-6000, 800-876-4766 (619 area code only)

California Poison Control System—San Francisco, San Francisco General Hospital, 1001 Potrero Avenue, Building 80, Room 230, San Francisco, CA 94110
Emergency Phone: 800-523-2222

California Poison Control System—Sacramento, Regional Poison Control Center, 2315 Stockton Boulevard, Room 1024, House Staff Facility, Sacramento, CA 95817
Emergency Phone: 916-734-3692, 800-342-9293 (Northern CA only)

COLORADO

Rocky Mountain Poison and Drug Center, 8802 E. 9th Avenue, Denver, CO 80220-6800
Emergency Phone: 303-629-1123, COLO WATTS 800-332-3073, MONT WATTS 800-525-5042, NEV WATTS 800-446-6179

CONNECTICUT

Connecticut Poison Control Center, University of Connecticut Health Center, 263 Farmington Avenue, Farmington, CT 06030
Emergency Phone: 800-343-2722 (CT Only), 203-679-3056

DELAWARE

The Poison Control Center, 3600 Sciences Center, Ste. 220, Philadelphia, PA 19104-2641

Emergency Phone: 215-386-2100, 800-722-7112

DISTRICT OF COLUMBIA

National Capitol Poison Center, 3201 New Mexico Avenue, N.W., Ste. 310, Washington, DC 20016

Emergency Phone: 202-625-3333

FLORIDA

Florida Poison Information Center–Jacksonville, University Medical Center, University of Florida Health Science Center/Jacksonville, 655 West 8th Street, Jacksonville, FL 32209

Emergency Phone: 800-282-3171 (FL only), 904-549-4465

Florida Poison Information Center–Miami, University of Miami/Jackson Memorial Hospital, 1611 NW 12th Avenue, Urgent Care Center Bldg., Rm. 219, Miami, FL 33136

Emergency Phone: 800-282-3171(FL Only)

Florida Poison Information Center and Toxicology Resource Center, Tampa General Hospital, PO Box 1289, Tampa, FL 33601

Emergency Phone: 813-256-4444 (Tampa), 800-282-3171 (FL only)

GEORGIA

Georgia Poison Center, Hughes Spalding Children's Hospital, Grady Health Systems, 80 Butler Street S.E. P.O. Box 26066, Atlanta, GA 30335-3801

Emergency Phone: 404-616-9000, 800-282-5846 (GA only)

HAWAII

Hawaii Poison Center, 1319 Punahou Street, Honolulu, HI 96826

Emergency Phone: 808-941-4411

IDAHO

Idaho Poison Center, 3092 Elder Street, Boise, ID 83720-0036

Emergency Phone: 208-334-4570, 800-632-8000 (ID Only)

ILLINOIS

Illinois Regional Poison Center, Rush-Presbyterian-St. Luke's Medical Center, 1653 West Congress Parkway, Chicago, IL 60612

Emergency Phone: 312-942-5969, 800-942-5969

INDIANA

Indiana Poison Center, Methodist Hospital of Indiana, I-65 & 21st Street, P.O. Box 1367, Indianapolis, IN 46206-1367

Emergency Phone: 317-929-2323, 800-382-9097 (IN only)

IOWA

St. Luke's Poison Center, St. Luke's Regional Medical Center, 2720 Stone Park Boulevard, Sioux City, IA 51104

Emergency Phone: 712-277-2222, 800-352-2222

Mid-Iowa Poison and Drug Information Center, Variety Club Poison and Drug Information Center, Iowa Methodist Medical Center, 1200 Pleasant Street, Des Moines, IA 50309

Emergency Phone: 515-241-6254, 800-362-2327 (IA only)

Poison Control Center, The University of Iowa Hospitals and Clinics, Pharmacy Department, 200 Hawkins Drive, Iowa City, IA 52242

Emergency Phone: 800-272-6477

KANSAS

Mid-America Poison Control Center, University of Kansas Medical Center, 3901 Rainbow Boulevard, Room B-400, Kansas City, KS 66160-7231

Emergency Phone: 913-588-6633, 800-332-6633 (KS only and KC metro area)

KENTUCKY

Kentucky Regional Poison Center of Kosair's Children's Hospital, Medical Towers South, Ste. 572, PO Box 35070, Louisville, KY 40232-5070

Emergency Phone: 502-589-8222, 800-722-5725 (KY only)

LOUISIANA

Louisiana Drug and Poison Information Center, Northeast Louisiana University, Sugar Hall, Monroe, LA 71209-6430

Emergency Phone: 800-256-9822 (LA only), 318-362-5393

MAINE

Maine Poison Control Center, Maine Medical Center, Department of Emergency Medicine, 22 Bramhall Street, Portland, ME 04102

Emergency Phone: 207-871-2950, 800-442-6305 (ME only)

MARYLAND

Maryland Poison Center, University of Maryland School of Pharmacy, 20 N. Pine Street, Baltimore, MD 21201

Emergency Phone: 410-528-7701, 800-492-2414 (MD only)

MASSACHUSETTS

Massachusetts Poison Control System, 300 Longwood Avenue, Boston, MA 02115

Emergency Phone: 617-232-2120, 800-682-9211

MICHIGAN

Blodgett Regional Poison Center, 1840 Wealthy S.E., Grand Rapids, MI 49506-2968

Emergency Phone: 800-POISON-1

Children's Hospital of Michigan Poison Control Center, Harper Professional Office Building, 4160 John R., Ste. 425, Detroit, MI 48201

Emergency Phone: 313-745-5711, 800-764-7661

MINNESOTA

Hennepin Regional Poison Center, Hennepin County Medical Center, 701 Park Avenue, Minneapolis, MN 55415

Emergency Phone: 612-347-3141, Petline: 612-337-7387, 612-337-7474

Minnesota Regional Poison Center, 8100 34th Avenue South, P.O. Box 1309, Minneapolis, MN 55440-1309

Emergency Phone: 612-221-2113

MISSISSIPPI

Mississippi Regional Poison Control Center, University of Mississippi Medical Center, 2500 North State Street, Jackson, MS 39216-4505

Emergency Phone: 601-354-7660

MISSOURI

Cardinal Glennon Children's Hospital, Regional Poison Center, 1465 S. Grand Boulevard, St. Louis, MO 63104

Emergency Phone: 314-772-5200, 800-366-8888, 800-392-9111

Children's Mercy Hospital, 2401 Gillham Road, Kansas City, MO 64108

Emergency Phone: 816-234-3430

MONTANA

Serviced by: Rocky Mountain Poison and Drug Center, 8802 E. 9th Avenue, Denver, CO 80220-6800

Emergency Phone: 303-629-1123, COLO WATTS 800-332-3073, MONT WATTS 800-525-5042, NEV WATTS 800-446-6179

NEBRASKA

The Poison Center, 8301 Dodge Street, Omaha, NE 68114

Emergency Phone: 402-390-5555 (Omaha), 800-955-9119 (NE &WY)

NEVADA

Serviced by: Rocky Mountain Poison and Drug Center, 8802 E. 9th Avenue, Denver, CO 80220-6800

Emergency Phone: 303-629-1123, COLO WATTS 800-332-3073, MONT WATTS 800-525-5042, NEV WATTS 800-446-6179

NEW HAMPSHIRE

New Hampshire Poison Information Center, Dartmouth-Hitchcock Medical Center, One Medical Center Drive, Lebanon, NH 03756

Emergency Phone: 603-650-8000, 603-650-5000(11pm-8am), 800-562-8236(NH only)

NEW JERSEY

New Jersey Poison Information and Education System, 201 Lyons Avenue, Newark, NJ 07112

Emergency Phone: 800-POISON1(800-764-7661)

NEW MEXICO

New Mexico Poison and Drug Information Center, University of New Mexico, Health Science Center Library, Rm. 125, Albuquerque, NM 87131-1076

Emergency Phone: 505-843-2551, 800-432-6866 (NM only)

NEW YORK

Central New York Poison Control Center, SUNY Health Science Center, 750 E. Adams Street, Syracuse, NY 13210

Emergency Phone: 315-476-4766, 800-252-5655

Finger Lakes Regional Poison Center, Box 777, University of Rochester Medical Center, 601 Elmwood Avenue, Box 321, Rm. G-3275, Rochester, NY 14642

Emergency Phone: 716-275-5151, 800-333-0542

Husdon Valley Poison Center, Phelps Memorial Hospital Center, 701 N. Broadway, North Tarrytown, NY 10591

Emergency Phone: 800-336-6997, 914-366-3030

Long Island Regional Poison Control Center, Winthrop University Hospital, 259 First Street, Mineola, NY 11501

Emergency Phone: 516-542-2323

New York City Poison Control Center, N.Y.C. Department of Health, 455 First Avenue, Room 123, New York, NY 10016
Emergency Phone: 212-340-4494, 212-POISONS

Western New York Regional Poison Control Center, Children's Hospital of Buffalo, 219 Bryant Street, Buffalo, NY 14222
Emergency Phone: 716-878-7654, 7655, 7856, 7857

NORTH CAROLINA

Carolinas Poison Center, 1012 S. Kings Drive, Ste. 206, P.O. Box 32861, Charlotte NC 28232-2861
Emergency Phone: 704-355-4000, 800-84-TOXIN(1-800-848-6946)

Triad Poison Center, 1200 N. Elm Street, Greensboro, NC 27401-1020
Emergency Phone: 910-574-8105, 800-953-4001 (NC only)

NORTH DAKOTA

North Dakota Poison Information Center, MeritCare Medical Center, 720 4th Street North, Fargo, ND 58122
Emergency Phone: 701-234-5575, 800-732-2200 (ND, MN, SD only)

OHIO

Akron Regional Poison Center, 1 Perkins Square, Akron, OH 44308
Emergency Phone: 216-379-8562, 800-362-9922 (OH only)

Central Ohio Poison Center, 700 Children's Drive, Columbus, OH 43205-2696
Emergency Phone: 614-228-1323, 800-682-7625

Cincinnati Drug & Poison Information, Center and Regional Poison Control System, P.O. Box 670144, Cincinnati, OH 45267-0144
Emergency Phone: 513-558-5111, 800-872-5111(OH only)

Greater Cleveland Poison Control Center, 11100 Euclid Avenue, Cleveland, OH 44106
Emergency Phone: (Toll-Free) 888-231-4455; 216-231-4455

Medical College of Ohio Poison and Drug Information Center, 3000 Arlington Avenue, Toledo, OH 43614

Emergency Phone: 419-381-3897, 800-589-3897(419 area code only)

Northeast Ohio Poison Education/Information Center, 1320 Timken Mercy Drive N.W., Canton, OH 44708

Emergency Phone: 800-456-8662 (OH only)

OKLAHOMA

Oklahoma Poison Control Center, 940 N.E. 13th Street, Rm. 3N118, Oklahoma, OK 73104

Emergency Phone: 405-271-5454, 800-522-4611(OK Only)

OREGON

Oregon Poison Center, Oregon Health Sciences University, CB550, 3181 S.W. Sam Jackson Park Road, Portland, OR 97201

Emergency Phone: 503-494-8968, 800-452-7165 (OR only)

PENNSYLVANIA

Central Pennsylvania Poison Center, University Hospital, Milton S. Hershey Medical Center, Hershey, PA 17033-0850

Emergency Phone: 800-521-6110, 717-531-6111

Lehigh Valley Hospital Poison Prevention Program, 17th & Chew Streets, P.O. Box 7017, Allentown, PA 18105-7017

Administration Phone: 610-402-2536

The Poison Control Center, 3600 Sciences Center, Ste. 220, Philadelphia, PA 19104-2641

Emergency Phone: 215-386-2100, 800-722-7112

Pittsburgh Poison Center, 3705 Fifth Avenue, Pittsburgh, PA 15213

Emergency Phone: 412-681-6669

Regional Poison Prevention Education Center, Mercy Regional Health System, 2500 Seventh Avenue, Altoona, PA 16602

Administrative Phone: 814-949-4197

RHODE ISLAND

Rhode Island Poison Center, 593 Eddy Street, Providence, RI 02903

Emergency Phone: 401-444-5727

SOUTH CAROLINA

Palmetto Poison Center, College of Pharmacy, University of South Carolina, Columbia, SC 29208

Emergency Phone: 803-765-7359, 800-922-1117 (SC only), 706-724-5050, 803-777-1117

SOUTH DAKOTA

McKennan Poison Control Center, Box 5045, 800 E. 21st Street, Sioux Falls, SD 57117-5045

Emergency Phone: 605-336-3894, 800-952-0123, 800-843-0505

TENNESSEE

Middle Tennessee Poison Center, The Center for Clinical Toxicology, Vanderbilt University Medical Center, 1161 21st Avenue South, 501 Oxford House, Nashville, TN 37232-4632

Emergency Phone: 615-936-2034 (local), 800-288-9999 (regional)

Southern Poison Center, Inc., 847 Monroe Avenue, Ste. 230, Memphis, TN 38163

Emergency Phone: 901-528-6048; 800-228-9999 (TN only)

TEXAS

Central Texas Poison Center, Scott & White Memorial Clinic & Hospital, 2401 S. 31st Street, Temple, TX 76508

Emergency Phone: 817-774-2005, 800-POISON1, 800-764-7661

North Texas Poison Center, Texas Poison Center Network at Parkland Memorial Hospital, 5201 Harry Hines Boulevard, P.O. Box 35926, Dallas, TX 75235

Emergency Phone: 800 POISON1 (800-746-7661)

South Texas Poison Center, University of Texas Health Science Center, 146 Forensic Science Center, San Antonio, TX 78284-7849

Emergency Phone: 800-POISON1 (800-764-7661) (TX Only)

Texas Poison Control Network at Amarillo, PO Box 1110, 1501 S. Coulter, Amarillo, TX 79175
Emergency Phone: 800-764-7661

Texas Poison Control Network at Galveston, Southeast Texas Poison Center, The University of Texas Medical Branch, 301 University Blvd., Galveston, TX 77555-1175
Emergency Phone: 409-765-1420 (Galveston), (713) 654-1701 (Houston) 800-764-7661 (TX Only)

West Texas Regional Poison Center, 4815 Alameda Avenue, El Paso, TX 79905
Emergency Phone: 800-764-7661 (TX Only)

UTAH

Utah Poison Control Center, 410 Chipeta Way, Ste. 230, Salt Lake City, UT 84108
Emergency Phone: 801-581-2151, 800-456-7707 (UT only)

VERMONT

Vermont Poison Center, Fletcher Allen Health Care, 111 Colchester Avenue, Burlington, VT 05401
Emergency Phone: 802-658-3456

VIRGINIA

Blue Ridge Poison Center, University of Virginia, Box 67, Blue Ridge Hospital, Charlottesville, VA 22901
Emergency Phone: 804-924-5543, 800-451-1428

Virginia Poison Center, 401 N. 12th Street, Virginia Commonwealth University, Richmond, VA 23298-0522
Emergency Phone: 804-828-9123 (Richmond), 800-552-6337 (VA only)

WASHINGTON

Washington Poison Center, 155 NE 100th St., Ste. 400, Seattle, WA 98125
Emergency Phone: 206-526-2121, 800-732-6985 (WA only)

WEST VIRGINIA

West Virginia Poison Center, 3110 MacCorkle Avenue S.E., Charleston, WV 25304

Emergency Phone: 304-348-4211, 800-642-3625 (WV only)

WISCONSIN

Children's Hospital of Wisconsin Poison Center, P.O. Box 1997, Milwaukee, WI 53201

Emergency Phone: 414-266-2222, 800-815-8855 (WI Only)

University of Wisconsin Hospital Regional Poison Center, E5/238 CSC, 600 Highland Avenue, Madison, WI 53792

Emergency Phone: 608-262-3702, 800-815-8855

WYOMING

Serviced by: The Poison Center, 8301 Dodge Street, Omaha, NE 68114

Emergency Phone: 402-390-5555 (Omaha), 800-955-9119 (NE &WY)

Canadian Poison Control Centers

Alberta

Poison and Drug Information Services, Foothills General Hospital, 1403-29th St. N.W. Calgary, AB T2N 2T9

1-800-332-1414; (403) 270-1414 local

British Columbia

B.C. Drug and Poison Information Centre, St. Paul's Hospital, 1081 Burrard St., Vancouver, BC V6Z 1Y6

1-800-567-8911; (604) 682-5050 local

Manitoba

Provincial Poison Information Centre, Children's Hospital Health Sciences Centre, 840 Sherbrook St., Winnipeg, MB R3A 1S1

(204) 787-2591 emergency inquiries; (204) 787-2444

New Brunswick

Poison Control Centre, The Moncton Hospital, 135 McBeath Ave., Moncton, NB E1C 6Z8

(506) 857-5555 emergency inquiries; (506) 857-5353

Emergency Department, Saint John Regional Hospital, P.O.Box 2100; Saint John, NB E2L 4L2

(506) 648-6222 local

Newfoundland

Provincial Poison Control Centre, The Dr. Charles A. Janeway Child Health Centre, 710 Janeway Place, St. John's, NF A1A 1R8

(709) 722-1110

Northwest Territories

Emergency Department, Stanton Yellowknife Hospital, P.O.Box 10, Yellowknife, NT X1A 2N1

(403) 920-4111 local

Nova Scotia

Poison Control Centre, The Izaak Walton Killam Children's Hospital, P.O.Box 3070, Halifax, NS B3J 3G9

(902) 428-8161 local; (902) 428-3213

Ontario

Ontario Regional Poison Information Centre, Children's Hospital of Eastern Ontario, 401 Smyth Road, Ottawa, ON K1H 8L1

1-800-267-1373 toll-free Ontario; (613) 737-1100 emergency inquiries; (613) 737-2320

Ontario Regional Poison Information Centre, The Hospital for Sick Children, 555 University Ave., Toronto, ON M5G 1X8;

1-800-268-9017 toll-free; (416) 598-5900 local

Prince Edward Island

Poison Control Centre, The Izaak Walton Killam Children's Hospital, P.O.Box 3070, Halifax, NS B3J 3G9

1-800-565-8161 toll free from P.E.I

Quebec

Centre antipoison du Québec, Le Centre Hospitalier de l'Université Laval 2705 Boul. Laurier, Sainte-Foy, PQ G1V 4G2

1-800-463-5060; (418) 656-8090

Saskatchewan

Emergency Department, Regina General Hospital, 1440 14th Ave., Regina, SK S4P 0W5

1-800-667-4545; (306) 359-4545

Emergency Department, Royal University Hospital, Saskatoon, SK S7N 0X0

1-800-363-7474; (306) 966-1010

Yukon Territory

Emergency Department, Whitehorse General Hospital, 5 Hospital Road, Whitehorse, YT Y1A 3H7

(403) 667-8726

INDEX